PHYSIOLOGY: CUSTOM-DESIGNED CHEMISTRY

GETTING PAST THE ANXIETY THAT PHYSIOLOGICAL CHEMISTRY IS TOO HARD TO LEARN

MARGARET THOMPSON REECE PHD

Copyright © 2012 by Margaret Thompson Reece. Registration Number TXu 1-842-025 issued by The United States Copyright Office effective December 12, 2012.

Figures 2, 3, 4, 6, 7, 10, 11, 15, 16, 18, 19, 20, 21 and 22 are licensed from Shutterstock.com, a stock photo company. Remaining figures, 1, 5, 8, 9, 12, 13, 14, and 17, are under a Wikimedia Commons license.

All rights reserved. No part of this publication may be reproduced, distributed or transmitted in any form or by any means, including photocopying, recording, or other electronic or mechanical methods, without prior written permission, except in the case of brief quotations embodied in critical reviews and certain other noncommercial uses permitted by copyright law. For permission requests, write to the address below.

Margaret Thompson Reece
Reece Biomedical Consulting LLC
8195 Cazenovia Road
Manlius, NY 13104
www.medicalsciencenavigator.com

Book Layout ©2013 BookDesignTemplates.com

Cover Photo © 2013 Yeko Photo Studio

Ordering Information:
Quantity sales: For details, contact DrReece@medicalsciencenavigator.com.

Physiology: Custom-Designed Chemistry/ Margaret Thompson Reece -- 1st ed.
ISBN 978-1-4823266-1-1

This book is dedicated to my many mentors at the University of California at Davis.

Science is a way of thinking much more than it is a body of knowledge.

–CARL SAGAN

PREFACE

Why learn chemistry in a physiology course? Some universities have removed physiology from their curriculum, because students and teachers felt it was too "hard" a course. However, entrance to healthcare professional training programs usually requires that students complete credits for some form of physiology course.

Having been a physiologist for over 30 years, and having watched my students struggle with this subject, I came to the conclusion that the key to understanding this science is to grasp how living systems creatively tailor a limited number of chemistry's principles for their own use.

Most anatomy and physiology courses start with a week of lectures focused upon basic chemistry. Students ask "Am I in the wrong class? I thought I signed up for physiology!" It feels rather like getting on an airplane to Washington DC and finding out by surprise that your flight is going to New York City.

The temptation is to skim over the chemistry section and move on to reading the next set of material about how cells function. The problem with doing this does not surface until later in the course when chemistry's principles are repeatedly used to support explanations of particular physiologic functions.

The expectation is that students clearly understand the underlying chemistry based upon the initial quick review. After all, there were few questions from the class during that lecture. Unfortunately, what really happens is that students struggle for the rest of the course to make connections between the principles of chemistry and specific theories of physiologic function.

Most college physiology courses for non-science majors lack a chemistry prerequisite. Students headed for careers in healthcare, who are required to pass physiology with a B or better grade, often have no previous education in chemistry.

Granted some students wisely take courses in chemistry before enrolling in physiology. But often even they struggle with the application of chemistry's principles to physiology. General chemistry courses do not adequately prepare life science students to understand fully the degree to which water is a highly reactive chemical. Nor do chemistry courses explore chemical reactions in a constant temperature environment.

This book and several more that I am writing for life science students are designed to provide you with a framework for understanding the custom-designed chemistry of physiology. I was taught the basics of this framework 30 years ago by a superb group of professors

PREFACE

at the University of California, Davis, and it has proven to be relatively timeless. New understanding of the details of physiology has progressed enormously since then, including sequencing of the human genome and the explosion of new knowledge about the human immune defense system. Twenty or thirty years from now the details will expand even further, but the basic principles will remain the same.

INTRODUCTION

Chapters in this book are designed to help you appreciate selected aspects of chemistry that are used to explain functions of living organisms. Each chapter concentrates upon a subset of chemistry's principles needed to navigate theories of physiological function.

Each chapter is organized to develop a clear understanding of the chemistry one-step-at-a-time. Please do not skip the sections where you think you already know the material. The sequence of the ideas is more valuable than the concepts in isolation.

Each topic is a milestone on a journey to comprehension of the language of physiological chemistry. The end goal of this book is for you to be able to automatically draw upon the relevant chemical models during physiology lectures.

For example, when your instructor uses chemical terms to describe muscle contraction, you will be able to focus exclusively on the unique properties of muscle cells. You will not have to dig deep in your memory to recover the meaning of the chemical terminology while missing the material about the role of calcium in the mechanics of muscle cell shortening.

PHYSIOLOGY: CUSTOM-DESIGNED CHEMISTRY

In Chapter 1, <u>The energy that sets physiological processes in motion</u>, you will learn about the prevalent misuse of the word "chemical". When you complete this chapter you will have a clear understanding of what atoms and molecules really are and why it matters that you know. This chapter also delves into explaining that mysterious thing called "molecular bond energy". At the end you will also grasp how living systems really extract energy from food. This chapter contains the foundation upon which the following chapters are built. So, please, please read Chapter 1 first.

Chapter 2, <u>Water – a powerfully energetic chemical</u> is about water's chemical nature. I will try to convince you that water is not just filler material that surrounds important molecules such as proteins, fat, DNA, carbohydrates etc. Also the concept of pH will be explored. pH can be a particularly hard subject to fully appreciate. Yet it is one of the most tightly regulated variables in humans. Aspects of how that regulation is chemically accomplished will be discussed.

In Chapter 3, <u>How molecules mingle and relocate – diffusion, osmosis, osmotic pressure, & hydrostatic pressure</u> the real movers and shakers of physiology are featured. The chemistry that allows molecules to move within the human body includes diffusion, osmotic pressure, and hydrostatic pressure. When teaching physiology, there is seldom sufficient time to expand upon the

INTRODUCTION

details of these processes. Yet, so much of physiology depends on students reflexively knowing what they are. Students need to be so familiar with these pressures that they automatically know what the words mean when the instructor uses them.

Chapter 4, <u>Physiology's interface with Earth's atmosphere – exploring gas laws</u> examines nature's laws of gas behavior that are encountered in introductory physiology courses. They include Boyle's Law, Charles's Law, Gay-Lussac's Law, the Combined Gas Law, Dalton's Law of Partial Pressures, and Henry's Law. We live in an air environment (21% oxygen, 78% nitrogen, and 1% carbon dioxide and other gases), and we have evolved to be totally dependent on this environment for our source of life giving energy.

Chapter 5, <u>Fluid compartments: the platform for long distance communication</u> describes how diffusion of molecules across cell membranes between the body's intracellular and extracellular fluid compartments facilitates communication over long distances within the body. The skeletal muscle motor neuron is used to illustrate this process.

This book is the first in a series of books that I am writing for students who struggle to understand physiology. It is often hard to know exactly where to begin with human physiology, because the body's systems are complex and interdependent.

PHYSIOLOGY: CUSTOM-DESIGNED CHEMISTRY

This book focuses upon water's chemistry, because without water's chemistry there can be no physiology. Future books will expand these principles to include physiology of the major cell types and the three major systems that the body uses for communication over long distances– the nervous system, the cardiovascular system, and the hormonal system.

TABLE OF CONTENTS

PREFACE — vii
INTRODUCTION — xi

CHAPTER 1: The Energy that Sets Physiological Processes in Motion — 1
CHAPTER 2: Water – a Powerfully Energetic Chemical — 17
CHAPTER 3: How Molecules Mingle and Relocate – Diffusion, Osmosis, Osmotic Pressure, & Hydro-Static Pressure — 41
CHAPTER 4: Physiology's Interface with Earth's Atmosphere – Exploring Gas Laws — 59
CHAPTER 5: Fluid Compartments: the Platform For Long Distance Communication — 75
KEYWORD INDEX — 99
ABOUT THE AUTHOR — 101

CHAPTER 1

THE ENERGY THAT SETS PHYSIOLOGICAL PROCESSES IN MOTION

This first chapter is about one of the hardest to pin down concepts in physiological chemistry – what energy really is. It is also very difficult for students new to physiology and biochemistry to capture the essence of how energy moves from one molecule to another. Yet your lectures on the subject of nutrition, for example, will rely greatly upon a basic understanding of chemical energy and its transfer from molecule to molecule. So, let's begin with deciding what a chemical is.

What is a chemical?

There is a lot of confusion about what constitutes a chemical. Even chemistry textbooks have been known to define chemical substances as "any material with a definite chemical composition." Dictionary definitions are not much better. For example, one popular dictionary says a chemical is "of or relating to chemistry." It is bad form to use the word you're defining in the definition of that word.

Much of what we hear in the popular press also misleads us about the definition of a chemical. "Chemical-free" is a term widely used in marketing to imply safety of a product. Chemical–free is used to convey that a product is a natural substance, and natural is good. We are left with the idea that there is something bad about chemicals and good about natural substances.

Actually, the word chemical is a synonym for matter. Matter makes up the universe. Matter has an established arrangement of atoms and each form of matter has characteristic properties of its own. Matter known as silver possesses different properties than matter known as gold.

That means, in reality, everything is a chemical – any liquid, solid, gas – be it synthetic or natural. Water is a chemical, sugar is a chemical, fat is a chemical, meat (muscle) is composed of chemicals, and gasoline is a mixture of chemicals. We are made of chemicals.

What are atoms?

For our purpose we can think of atoms as being made up of three components, electrons, protons, and neutrons. The number of protons that an atom has determines the unique nature of the matter formed by that atom. For example, the smallest particle of carbon matter is an atom with 6 protons. Atoms making up lead matter have 82 protons, and atoms of gold matter have 79 protons.

The nucleus, or central part, of each atom includes protons that are positively charged and neutrons which have no electrical charge. Orbiting the nucleus much like the planets orbit the sun are the negatively charged electrons. Every atom has a number of electrons that is equal to its number of protons.

Electrons move around the nucleus so rapidly that it is not physically possible to tell where any one electron is at any given point in time. To get around the issue of electron location, scientists have agreed to predict the movement of electrons based upon mathematical probability equations.

Using probability estimates, it is now generally accepted that electrons circle their nucleus in groups that form orbital layers of various shapes.

For clarity the orbital's five components are displayed separately in Figure 1. A composite of the five components is displayed at the center of the bottom row.

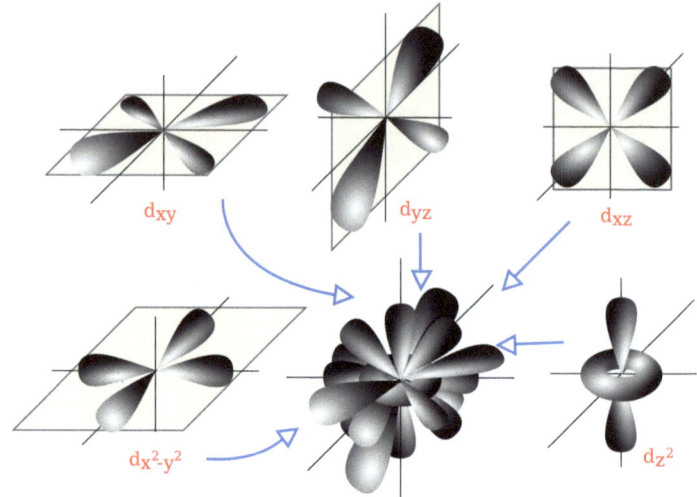

Figure 1: An overview of one type of electron orbital layer predicted by the probability equations, the d orbital. Photo © 2006 Sven

The d orbital displays some of the various shapes that electron orbital layers can assume. These shapes are based entirely on likelihood estimates of the actual location of electrons.

Each type of electron orbital layer is predicted to contain an optimal number of electrons. Electrons in the orbital layers located closest to a nucleus are most tightly held by the oppositely charged protons. Electrons in the orbital layer positioned farthest from a nucleus – the outermost orbital layer – are less tightly held and are sometimes able to form hybrid orbital layers with electrons of other atoms.

For most atoms their outermost orbital layer has more or less electrons than is optimal. Sharing of electrons by two or more atoms is a common method for optimizing the number of electrons in the outermost orbital layer of both. For example, carbon and lead each have 4 electrons in their outermost orbital layer available for sharing. Gold has only 1 electron to share. All orbital layers containing an optimal number of electrons are said to be "complete".

Links called *bonds* occur between two atoms when both atoms' outermost electron orbital becomes complete by the formation of a hybrid orbital. A pair of atoms can share one, two, or three electrons thus forming what chemists call single, double, and triple bonds. Or, a single atom such as carbon can share one of its 4 available electrons with each of 4 other atoms that need 1 more electron to complete their outermost electron orbital layer.

How do atoms become molecules?

If two atoms happen to meet as they move in space and if they have an arrangement of electrons such that they can form a hybrid orbital layer, then a chemical bond will form spontaneously between them.

Atoms connected by chemical bonds are called molecules. Chemical bonds are generally divided into two types, *covalent bonds* and *ionic bonds*. A covalent bond is a strong equal sharing of electrons by two nuclei.

Ionic bonds occur between two atoms that have greatly different size nuclei. An atom with a large very positive nucleus shares electrons very unequally with an atom that has a small, less positive nucleus. The larger nucleus almost steals the shared electron(s) away from the smaller atom. Ionic bonds are more easily broken than covalent bonds.

Where do molecules get their energy?

For this discussion we will confine ourselves to chemistry's definition of energy and leave the physics definition for another time. *Chemistry maintains that molecules possess kinetic energy due to their motion.* The chemical energy of molecules can, therefore, be experimentally determined by measuring the effect of that motion on some gauge.

Why molecules move, how they got started in motion, and why they do not stop are all theoretical questions without final answers. That molecules DO possess a predictable amount of motion (kinetic energy) at any given temperature is based on scientific observation. It is atomic motion, defined as kinetic energy, which humans and other life forms organize into special functions called physiology.

It has been experimentally demonstrated that the higher the temperature of a closed system, the faster the molecules in that system move. The human body operates within a very narrow temperature range – usually between 36.7°C and 37.2°C. This means that energy changes within the system that is the human body can be evaluated without accounting for large energy changes due to temperature fluctuation.

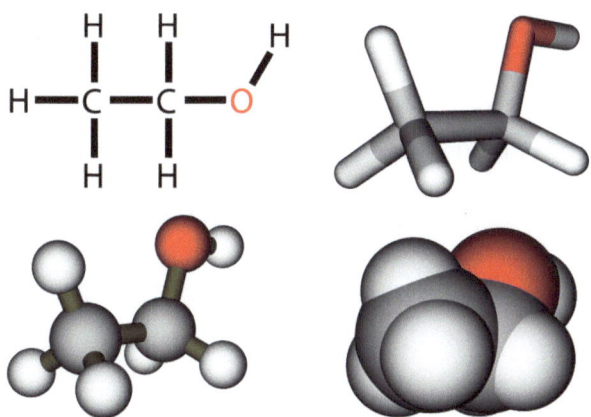

Figure 2: A representation of four ways in which chemists draw chemical bonds between ethyl alcohol atoms. Photo © 2012 Leonid Andronov.

The inherent property of molecules to move is often not emphasized in physiology textbooks. Rather molecules are drawn in one of several static formats. Such shorthand, as is shown in Figure 2, is useful but misleading. When students study reactions that build and break apart bonds in molecules drawn in this manner, the role kinetic energy plays is not always obvious.

Motion at the molecular level in humans is not noticeable. It is difficult to imagine, because our bodies appear to be relatively solid objects. The thought of all that random motion under the surface is a bit alien. So, equating the energy we use to move our solid appearing bodies with kinetic energy of our body's molecules requires some innovative thinking.

While we are being creative in our thinking, we must also accommodate the fact that basic chemistry requires some custom tailoring to fit as an explanation of physiologic function. For example, chemists rely heavily on changes in temperature to achieve their desired outcomes. In contrast, humans require a form of chemistry designed to operate at a single, constant temperature.

How does energy transfer between molecules in a constant temperature environment?

Remember from the discussion above that covalent and ionic bonds form spontaneously between atoms if it is possible for their outermost orbital layers to mingle. It is a matter of the right partners with sufficient kinetic energy coming together in the correct orientation. The more kinetic energy an atom has, the faster it moves and the more likely it will meet a possible partner in the correct orientation.

A *low energy bond* between atoms is characterized by a strong influence of both nuclei on the shared electron orbital layer. The "low energy" description found in chemistry text books comes from the fact that bound atoms have lower kinetic energy, or freedom of motion, than before the bond was formed. Atoms linked together by a hybrid orbital layer move more slowly than they did as separate entities. The phrase "strong bond" is generally used interchangeably with the description "low energy bond".

As mentioned above, chemists describe covalent bonds in terms of how many electrons are shared by the two molecules. In a single bond, one electron is shared, a double bond two electrons, and a triple bond three electrons. Usually, a single covalent bond between two atoms is weaker, that is the atoms retain higher kinetic energy, than a double bond, which in turn is weaker than a triple bond.

Exceptions to the rule, however, include hydrogen atoms that form relatively strong (low energy) single bonds with carbon, nitrogen, and oxygen atoms. Like-

PHYSIOLOGY: CUSTOM-DESIGNED CHEMISTRY

wise carbon forms strong single bonds with itself and with hydrogen, nitrogen, and oxygen. Carbon is unusual in its ability to form strong bonds with itself. Single bonds between other like atoms are relatively "weak" bonds or "high energy" bonds.

For ionic bonds, where two atoms have greatly different sized nuclei, the freedom of movement of the larger atom is diminished only a little and the smaller atom is along for the ride. These are said to be "weak bonds" with "high energy", because both atoms have retained a fairly large portion of their original kinetic energy. Such bonds are also readily broken when in the presence of molecules competing for the shared electron like water molecules. Solution of ionic compounds in water will be discussed further in Chapter 2.

Living creatures have specialized systems for releasing energy contained in the covalent bonds of food molecules, sugars and fats, and for using that energy to create their own structure, to maintain their internal micro-environments, and to become mobile. Energy swaps between ingested sugar and fat and the body's structural and functional molecules is a highly organized set of processes in humans commonly referred to as metabolic pathways.

To come to terms with energy transfer between molecules of the metabolic pathways one must first accept as true a natural law named the *First Law of Thermody-*

namics. The First Law of Thermodynamics states that: *Energy can be transformed from one form to another, but cannot be created or destroyed.* This law is also sometimes known as the law of conservation of energy.

Applying this law, if a bond in a molecule is broken and then the exact same bond reforms, it is assumed that the amount of energy required to break the bond is exactly the same as the amount of energy released when the bond reformed. That is the amount of motion necessary to pull the atoms far enough apart that the hybrid orbital cannot be maintained equals the amount of motion acquired by neighboring molecules when the hybrid orbital formed.

Remember from the above discussion that when a bond forms between two atoms, the atoms loose some of their kinetic energy – that is, the movement of both atoms is restricted. The First Law of Thermodynamics tells us that their kinetic energy is not lost to the universe. Rather it is transferred to something else. So, in physiology when one set of atoms "lose" kinetic energy (less motion) by forming a covalent bond the nearby atoms "gain" kinetic energy (greater motion).

This model can be made intuitively reasonable. An example may be to think of biologic molecules that exist in close confinement in the human body to be similar to couples moving to music on a very overcrowded dance floor. If half of the couples (or molecules) should decide

PHYSIOLOGY: CUSTOM-DESIGNED CHEMISTRY

to move closer together and dance more slowly, the remaining couples (or molecules) will gain space and they will dance further apart and more energetically.

How then do living creatures profit from the energy contained in the bonds of food molecules such as sugar? Key to answering this question is that the bonds in the sugar molecules are relatively "weak" or "high energy" compared to the bonds in water and in carbon dioxide which are "strong" and "low energy". Living systems break apart sugar molecules ($C_6H_{12}O_6$) and build water (H_2O) and carbon dioxide (CO_2) from the component atoms and additional oxygen. It takes many small intermediate steps to actually accomplish this, but the overall reaction is:

$C_6H_{12}O_6$ (sugar) + 6 O_2 ⟶ 6 CO_2 + 6 H_2O + 2880 kJ (energy) per mole

A much smaller amount of energy is required to break the bonds in sugar than is released when the strong low energy covalent bonds of water and carbon dioxide are formed. The net excess of energy released is either used to break apart other bonds or to provide heat which locally increases the kinetic energy of nearby molecules favoring spontaneous bond formation.

In a physiological system molecular bonds are broken and new bonds formed constantly with the help of large specialized molecules called enzymes. The process of making and breaking bonds in patterned sequences allows kinetic energy to be dispersed among physiologic molecules in a controlled fashion.

How do enzymes provide the energy needed to break molecular bonds?

When molecules in motion collide their bonds are sometimes stretched and/or bent enough to break. If the molecules are moving too slowly (have low kinetic energy), or if they collide in a less than optimal orientation, they will simply bounce off each other and their bonds will remain intact.

The minimum energy necessary to achieve the correct velocity with enough molecules in a proper orientation to bend and break covalent bonds is called the *activation energy of a chemical reaction*. Methods used in chemistry, addition of external forms of energy such as heat, light, electricity and pressure, to achieve a sufficient activation energy to break chemical bonds are not useful to living organisms that maintain a rather low and constant temperature. Explosions are not permitted in physiology!

To solve the problem, living organisms create tiny environments on the surface of large molecules where diverse smaller molecules congregate in very close association with each other. This closeness of pulsating molecules aligned in proper collision orientation, stretches and bends selected covalent bonds sufficiently that they actually break in a low kinetic energy setting. This process of precise alignment of congregated molecules lowers the minimum energy needed to break molecular bonds – that is it lowers the activation energy of the chemical reaction.

Therefore, getting a proper collision orientation in living systems is not left to random chance as it is in a test tube. The large molecules that have a pocket on their electronic surface to attract sets of smaller molecules (reactants) are called enzymes. Within enzyme pockets the visiting molecules self-arrange guided by their own electronic nature into a proper collision orientation. Usually arrangement of the visiting molecules takes place in multiple steps with each step stretching and bending a little further the bonds that are meant to break. Eventually properly positioned stretched bonds break, and energy is released to the local environment. New bonds that are possible between atoms in the vicinity are then free to form spontaneously.

Summary

The major points that you will need to remember from this chapter about molecular energy are:

- Any liquid, gas, or solid, be it synthetic or natural, is a chemical
- The principles of chemistry maintain that molecules possess kinetic energy due to their motion
- Bonds, or links, occur between two atoms when both atoms' outermost electron orbital layer becomes complete by formation of a hybrid orbital layer
- Covalent and ionic bonds *form spontaneously* between atoms if it is possible for their outermost orbital layers to mingle
- Kinetic energy (the motion) of individual atoms is reduced when bonds form
- A "strong bond" is a "low energy" bond. A "weak bond" is a "high energy" bond
- When one set of atoms "lose" kinetic energy by forming a bond, the other atoms in the system "gain" an equal amount of kinetic energy
- When a set of reactions take place where weak bonds are broken and strong bonds are

formed, there is a net release of energy to the system. This is because breaking weak bonds takes less input of energy than is released by the formation of strong bonds.

❖ Within enzyme pockets, reactant molecules self-arrange guided by their own electronic nature into a proper collision orientation.

❖ Once oriented properly on the enzyme, molecular bonds of the reactant molecules are stretched and bent in a stepwise process until they break.

CHAPTER 2

WATER – A POWERFULLY ENERGETIC CHEMICAL

The purpose of this chapter is to create a narrative about water's role in physiology. It will construct a logical progression of ideas to help you better understand the immense importance of water to living beings. It will also include a few particulars about water that you may find surprising.

Water, because it is so familiar, is often viewed by students in physiology as space filler between functional biologic molecules such as proteins, fat, DNA, carbohydrates, etc. Nothing could be less true!

Water's chemistry defines physiology. No other chemical has the necessary properties to take the body's functional molecules into solution, and to facilitate their dynamic co-operation to produce life.

The chemistry of water continues to be an active area of scientific research. Scientists are still discovering unique aspects of water's molecular nature when it is confined to biologic spaces. Data suggest that some water molecules, those present in the very small pockets of large biologic molecules such as proteins, display an unusual atomic structure. Such data will cause some aspects of physiology to be re-appraised in the near future.

In the meantime, to make sense of water's role in larger physiologic spaces, there are two features of its nature that you will need to understand in detail. The first feature is the *polarity* of the water molecule [H_2O]. A water molecule's geometric structure has distinct opposite ends. One end carries a *positive partial electrical charge*, and the other a *negative partial electrical charge* because of unequal sharing of electrons between the atoms.

The second feature you need to understand is *pH*. The concept of pH originates from evidence that a small fraction of water molecules auto-dissociate when water is in its liquid form. That is, one of the hydrogen nuclei, a single proton, transfers to an adjacent water molecule forming a hydronium ion [H_3O^+]. The remnant water molecule now carries a full negative charge and is called a hydroxide ion [OH^-]. This is summarized by the equation below.

$$2\ H_2O \longleftrightarrow [H_3O^+] + [OH^-]$$

WATER – A CHEMICAL

Electrically charged molecules where the number of electrons is not equal to the number of protons are called *ions*. (There is additional discussion of atomic structure in Chapter 1) If ions have more protons than electrons – like [H_3O^+] – they are called *cations*. If ions have more electrons than protons – like [OH^-] – they are *anions*.

All except newest textbooks present the hydrogen cation as [H^+] rather than [H_3O^+] ignoring its complex interaction with other water molecules. However, this should not be a problem, because the body's regulatory mechanisms are explained equally well with either notation at the level of introductory physiology.

In pure water – which is seldom found – there is an equal number of hydronium and hydroxide ions. When these two ions are equal, pH is said to be neutral. Neutral pH is required for proper function of most biologic molecules. However, the body's energy capturing systems within cells introduce additional [H^+] to body fluids. The excess hydrogen ion causes pH to deviate from neutral pH 7 unless controls are put into place. The most widely used control system will be discussed later in this chapter.

These two features of water's chemistry, *molecular polarity* and *pH*, provide the infrastructure on which biologic systems are assembled. Becoming comfortable with these principles will greatly improve your understanding of physiology.

What is a 'partial' electrical charge?

Water's narrative begins with the creation of a water molecule [H_2O] when an atom of oxygen enters into covalent bonds with two atoms of hydrogen. (Refer back to Chapter 1 for more discussion about atoms entering into covalent bonds.)

To make a water molecule the electron orbital layers of two atoms of hydrogen and one atom of oxygen must be shared. But, sharing of electrons in this molecule is not equal between hydrogen with 1 positive proton and oxygen with 8 positive protons.

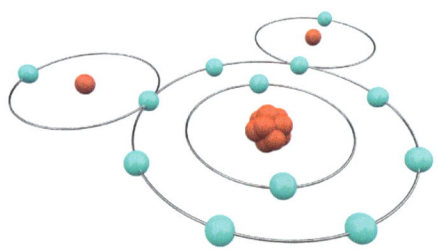

Figure 3: A schematic drawing of water's atomic structure showing electrons in each orbital layer. Photo © 2012 iQoncept

WATER – A CHEMICAL

Oxygen has a complete inner electron orbital layer with 2 electrons, and an outer orbital layer with 6 electrons. However, probability equations (for more on probability equations see Chapter 1) predict that the outer electron orbital layer of oxygen needs 8 electrons to be complete. To acquire a stable state with a complete outer electron orbital layer, oxygen shares its outer orbital layer with the 1 electron of two hydrogen atoms. This sharing creates two covalent bonds and one molecule of water.

Even though water is a mostly stable molecule, its electrons are not equally shared between oxygen and hydrogen. Remember that electrons are held in their orbits around an atomic nucleus by the pull of the positive charge of the protons in that nucleus. Oxygen's 8 protons create a much greater positive force on water's electrons than hydrogen's 2 protons.

Oxygen pulls hydrogen's electrons far away from the hydrogen nuclei. This creates a *partial negative charge* around the oxygen atom, due to the prolonged presence of extra electrons. It is not a full negative charge because the hydrogen nuclei retain a partial hold on their electrons. However, the hydrogen nuclei lack their normal level of negative charge and therefore possess a *partial positive charge*.

Scientists have concluded that unequal sharing of electrons by the atoms of water produces an approximate tetrahedral geometry for individual molecules.

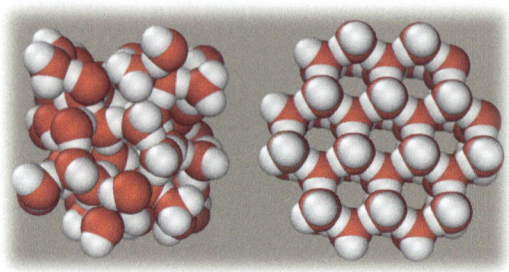

Figure 4: Experimentally determined bond angles of water molecules when they are in crystal form, ice. Photo © 2010 P99am

The polar electronic nature of water molecules allows them to form weak bonds with each other. Such bonds are different than either covalent or ionic bonds discussed in Chapter 1. The weak bonds that form between water molecules are called *hydrogen bonds*. When two water molecules in motion approach each other an electrostatic attraction occurs. The partial negative charge on the oxygen atom of the one water molecule is attracted to the partial positive charge on a hydrogen atom of the adjacent water molecule.

In its liquid form, closely packed water molecules possess a large amount of kinetic energy and thus move rapidly in random patterns. With all this motion hydrogen bonds between water molecules are made, broken quickly, and remade.

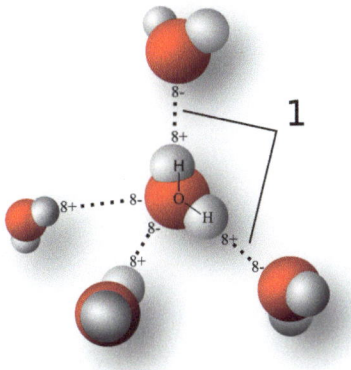

Figure 5: A model of liquid water molecules forming hydrogen bonds, the dashed lines, with each other. Photo © 2011 Qwerter

The kinetic energy of the large number of molecules per volume of liquid water is huge. It takes a substantial amount of additional heat to increase water's overall kinetic energy. This means that small fluctuations of heat energy due to molecular restructuring in the body change the temperature of the body's water phase insignificantly. It is this aspect of water that creates a heat sink to maintain human body temperature within a narrow range.

Similar unequal sharing of electrons between atoms also occurs in parts of other, larger biologic molecules.

This leads to partially positive and partially negative regions on the electronic surface of many molecules that can form hydrogen bonds with each other and with water molecules.

Why do only some molecules dissolve in water?

First we should talk about the relationship between hydrogen bonding of water molecules and the surface tension of water. It may seem that we are going in the wrong direction with this topic. Yet, this property of water really will bring us to an explanation of why only some molecules dissolve in water.

Surface tension develops at the air interface of a container of water. Such surface rigidity of water is sufficiently strong that some insects can walk on it – such as the *water strider* pictured on the next page.

Surface molecules of water at an air boundary are able to hydrogen bond with other water molecules neighboring them in the plane of the surface, and with molecules below them. There are no molecules above to balance forces from the molecules below. This tends to draw surface molecules down into bulk water. The net effect of these forces requires the plane of water's surface to contract to a minimal area that is somewhat firm.

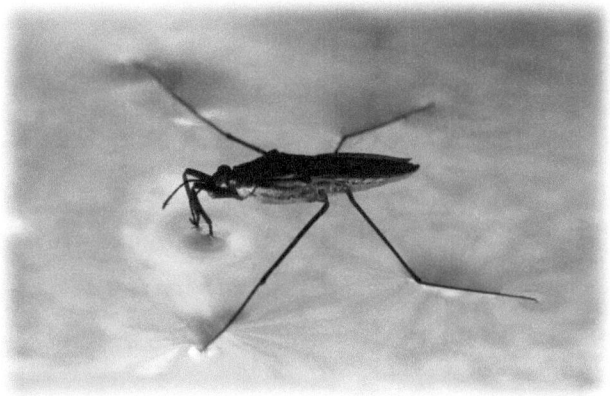

Figure 6: A water strider *Aquarius remigis* poised on the surface of pond water. Photo © 2012 BMJ

It is this same surface tension that causes small packets of water to form droplets on some objects. The geometric shape that has the smallest ratio of surface area to volume is the sphere. Droplets that form on materials lacking an electrical charge to attract the dipoles of water molecules form as spheres.

When an object that possesses electrical charges on its surface is coated with water, then water will spread on the surface as a uniform film and droplets will not form. Water spreading into a uniform film on a surface is called wetting. Wetting can be demonstrated with very clean glass. Glass is made of silicates that have negative charges that attract the hydrogen end of water molecules.

Similarly biologic molecules with a charge on their surface will interact with the dipole charges of water molecules and become wetted by water. Biologic molecules gain their partial charges in the same way water does. There is dipole formation in many biologic molecules due to unequal sharing of electrons in parts of the molecule. When such wetting interactions take place, biologic molecules are said to go into solution or dissolve in water. Molecules with charged surfaces that can be dissolved by water are described as hydrophilic, or 'water loving'. Biologic molecules known as proteins often belong to the hydrophilic group.

Figure 7: Near spherical water droplets on the waxy surface of a leaf. Photo © 2012 Tischenko Irina

Some biologic molecules do not exhibit dipole charges. As a result water cannot wet their surface and bring them into solution. Such molecules are said to be hydrophobic, or to have a 'water aversion'. Hydrophobic molecules include the class of molecules known as fats or lipids.

In living systems there are advantages to having both hydrophilic and hydrophobic molecules. Hydrophobic molecules partition off water compartments and provide other infrastructure that is part of human body anatomy. The primary role of hydrophilic molecules is to co-ordinate body functions such as extraction of energy from food.

How is human fluid similar to sea water?

Evolutionary theory maintains that the sea is where life began. This idea is strengthened by the fact that human body fluid is similar to sea water. Both are basically a salt solution composed of water, dissolved salts, and biologic molecules.

Salts are a large and important group of hydrophilic molecules. Salts are composed of atoms joined together by ionic bonds. In chapter 1 we learned that ionic bonds form between two atoms that have greatly different size nuclei. A major component of body water and sea water is the salt sodium chloride, NaCl. Albeit, the level of NaCl in sea water is much higher than it is in human body water.

The outermost electron orbital layer of chlorine only needs one electron to become complete. Sodium has only one lonely electron in its outermost orbital layer.

When these two atoms get close enough together they will form a spontaneous ionic bond that is weak and relatively high energy.

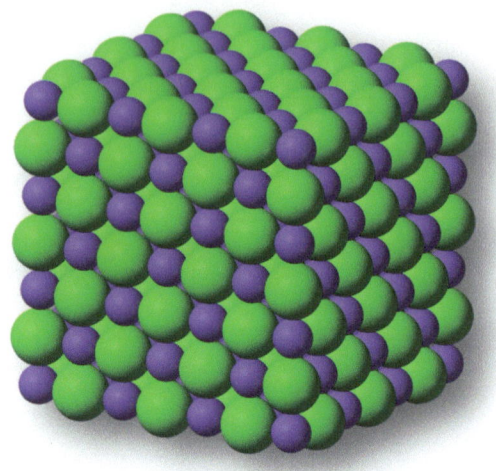

Figure 8: A model of a crystal of table salt, NaCl, which is held together by ionic bonds. Photo © 2009 Raj6

Sodium [Na – smaller purple spheres] has 11 protons compared to chlorine's [Cl – larger green spheres] 17 protons.

Ionic bonds such as those of crystalline NaCl are easily broken by water. Sodium's single electron is pulled so far toward the chlorine nucleus that it is relatively easy for the oxygen end of water molecules to form a cluster around the sodium nucleus repulsing its electron. Sodium's displaced electron then becomes firmly integrated into chlorine's orbit.

When sodium is separated from chlorine by water, chlorine becomes negatively charged, because it then has one more electron in its orbits than protons in its nucleus. The positive poles of water molecules are attracted to chlorine's negative charge. Quickly, chlorine is also at the center of a cluster of water molecules that prevent reformation of NaCl.

How is hydronium ion measured in fluid?

Human body fluid is primarily water and sodium chloride with a myriad of biologic molecules floating in it such as sugar, proteins, hormones etc. As the body processes food and uses energy extracted from food, hydrogen cations [H^+] are often transferred to water molecules forming hydronium ions [H_3O^+] in excess of those due to water alone.

So why do we care how many [H_3O^+] float around in body fluids bathing tissues and in blood? The primary reason is that they are likely to randomly break any chemical bond that they get near. When a bond breaks in large biologic molecules, it causes a change in molecular shape. Shape is very important to biologic molecules. When their shape alters, they can no longer get their job done.

PHYSIOLOGY: CUSTOM-DESIGNED CHEMISTRY

How many $[H_3O^+]$ are too many? We stated earlier in this chapter that even pure water has a low amount of $[H_3O^+]$. As it turns out the amount of $[H_3O^+]$ in pure water is close to optimal for humans. Much more than that amount causes problems.

Before we put a number to the optimal quantity of $[H_3O^+]$ in body fluids we need to recognize two things. First, not all body compartments want or need the same amount of $[H_3O^+]$ floating around. There are places like the stomach where breaking bonds in incoming food molecules is a good thing. There are other places such as at the small intestine where it is not a good thing for $[H_3O^+]$ to be damaging the transport molecules that take nutrients from the intestine into blood vessels.

Second, to quantify $[H_3O^+]$ a unit of measurement must be defined. To understand the unit used to quantify $[H_3O^+]$ in water solutions, we must first talk about some units of measurement used by chemists – *moles* and *molar*.

The International System of Units for Chemistry includes a quantity called a *mole*. A mole is defined to be 6.02×10^{23} molecules of a substance. For example, one mole of NaCl and one mole of H_2O each contain 6.02×10^{23} molecules. However, the molecular weight of NaCl is greater than the molecular weight of H_2O. So, 6.02×10^{23} molecules, or a mole, of NaCl weighs more than a mole of H_2O.

The mass (or weight) of a mole of NaCl equals the molecular weight of sodium (22.99) plus the molecular weight of chlorine (35.45) expressed in grams. Therefore, a mole of NaCl weighs 58.44 grams. Likewise the mass of a mole of H_2O equals the molecular weight of hydrogen (1.00) times 2 plus that of oxygen (16.00), or 18 grams.

To find the molecular weight of an atom you need to look it up on a Periodic Chart. There are several nicely done Periodic Charts online. Go to Google and type in "interactive periodic chart" or http://www.ptable.com.

Molar is used to designate how many moles (or what part of a mole) are present in a liter of liquid. Metric quantities are used for parts of a mole (1.0, 0.1, 0.01, 0.001, etc), because it is easy to count the number of places to the right of a decimal point to describe very small quantities. For example, 0.001 mole can also be expressed as 1×10^{-3} mole. The liquid quantities used for molar are all metric measures also, for example a liter, milliliter (1×10^{-3} liter), microliter (1×10^{-6} liter), etc. Most biologic molecules circulate in blood in the range of 1×10^{-3} to 1×10^{-12} mole/liter.

It has been determined experimentally that pure water has 0.0000001 mole of $[H_3O^+]$ in every liter. Or, it could be said that the concentration of $[H_3O^+]$ in pure water is 1×10^{-7} moles/liter. The logarithm of 10 in this expression, -7, is obtained by again counting the number of places to the right of the decimal point in the expression 0.0000001 mole.

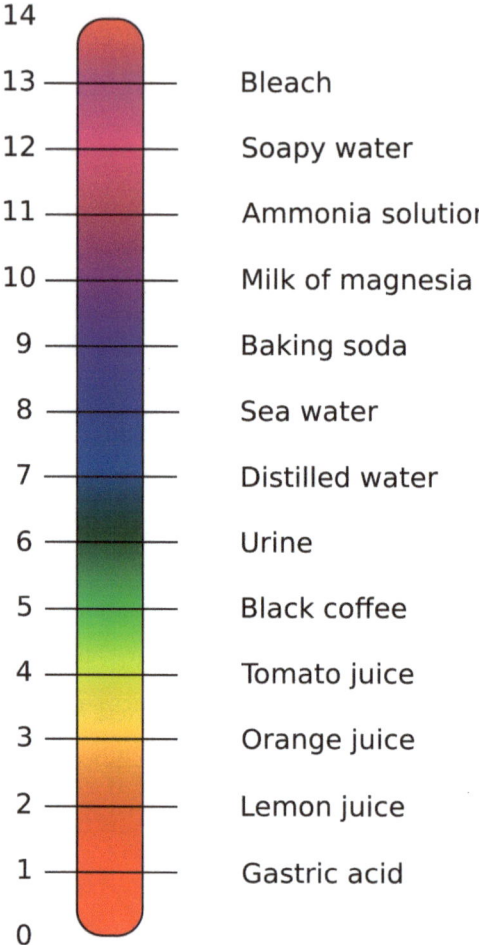

Figure 9: Scale showing pH value of some common liquids. Photo © Edward Stevens

The term pH is a shorthand value used to describe the concentration of [H_3O^+] in a water solution. pH is defined as the negative logarithm of the hydronium (or hydrogen) ion concentration. The definition of pH as a negative logarithm automatically converts the pH of pure water to a positive number, 7.

While the definition of pH appears to eliminate the problem of thinking in negative numbers, it really does not help that much. As the pH of a solution decreases from 7 to 3, [H_3O^+] increases in solution. That is because 1×10^{-3} is a much bigger number than 1×10^{-7}. In other words you still need to think of pH in terms of what negative logarithms of 10 signify.

Pictured on the previous page is a bar displaying the range of pH values from 1 to 14 of an array of common liquids. Comparing the value shown for urine, pH 6, and gastric acid, pH 1, it becomes clear that compartments within the human body can vary greatly in their concentration of [H_3O^+]. However, most body fluids such as blood and the fluid bathing cells are maintained near pH 7.

Refer back to the equation from above that summarizes [H_3O^+] formation in pure water. Notice that this equation shows that 2 water molecules produce an equal number of [H_3O^+] and [OH^-].

$$2\ H_2O \longleftrightarrow [H_3O^+] + [OH^-]$$

In biologic systems where [H_3O^+] and [OH^-] ions are constantly being added to body fluids during the breaking and making of molecular bonds, the quantity of hydronium and hydroxide ions are often not equal. Such molecular processing causes pH of the body fluid to shift in value.

When [H_3O^+] and [OH^-] are quantitatively equal in solution, that is at pH 7, the solution is said to be *neutral*. At pH values below 7, the solution is said to be *acidic*, that is it has more [H_3O^+] than [OH^-]. In contrast, at pH values larger than 7, the solution is said to be *alkaline* because it has more [OH^-] than [H_3O^+].

How does the human body regulate blood pH?

The normal pH range for human blood is 7.35 – 7.45. A blood pH below 7.35 is a condition known as acidosis and can produce coma in humans. A blood pH above 7.45 is called alkalosis and can cause uncontrollable muscle contraction.

Removal of excess [H_3O^+] from the human body is one of the functions of the kidneys. You will notice in the bar graph above that urine formed by the kidneys is more acidic than distilled water. Like the anatomy of the stomach the anatomy of the urinary bladder had been modified to protect it from damage by [H_3O^+]. The kidney is relatively slow in removing [H_3O^+], so other more immediate measures are also necessary to regulate blood pH.

The body's immediate and local response to pH deviation from neutral is managed by a mixture of molecules in blood called *buffers*. These mixtures are part weak acid and part base. An acid is any molecule soluble in water that breaks apart upon solution such that one fragment is an [H^+]. The other fragment formed upon dissociation of an acid carries a negative charge similar to the [OH^-] fragment of auto dissociated water. *The negatively charged fragment of a dissociated acidic molecule is called a base.*

There are many buffers used by the human body, but the most important one for maintaining neutral pH in blood is the carbonic acid – bicarbonate mixture. Carbonic acid has the molecular formula H_2CO_3. Bicarbonate, a base, has the molecular formula HCO_3^-. A summary of reactions available to carbonic acid are shown by the reaction sequences below.

$CO_2 + H_2O \longleftrightarrow H_2CO_3$ (carbonic acid)
$H_2CO_3 \longleftrightarrow HCO_3^-$ (bicarbonate) + H^+ (hydrogen ion)

Arrows pointing forward and backward indicate that these reactions can proceed in either direction. The hydrogen proton released by carbonic acid, one part of the buffering mixture, may combine with water to form $[H_3O^+]$ making the solution more acidic. Or, it may attach itself to a basic molecule – either its own base fragment bicarbonate, or another basic molecule present in solution, or with $[OH^-]$.

The second part of this buffering mixture, the base bicarbonate, is present in blood in greater quantity than carbonic acid. HCO_3^- in solution in excess to that formed by the dissociation of carbonic acid comes from other compounds such as sodium bicarbonate, $NaHCO_3$. Sodium bicarbonate is a salt, and it is dissolved in water much like NaCl.

This buffer mixture is very efficient in resisting changes in pH, because it is present in solution at a very high concentration relative to the $[H_3O^+]$ produced by molecular processing in the body's cells. To protect blood from becoming too basic, excess $[OH^-]$ produced by cell activity is combined with $[H_3O^+]$ from carbonic acid to form water. Excess $[H_3O^+]$ produced by cell activity such as contracting muscle combines with blood bicarbonate. This reaction protects against blood becoming too acidic.

The first of the two chemical equilibrium reactions shown above displays another valuable characteristic of the carbonic acid buffering system. Carbonic acid can be reversibly converted to carbon dioxide and water. It is that part of the reaction sequence that makes this buffering system particularly flexible. Because all these reactions are reversible, removing CO_2 from the left side of the equation – as when CO_2 is released from the lung – causes reformation of H_2CO_3 from HCO_3^- and H^+ reducing the level of acid $[H_3O^+]$ in blood.

Increasing CO_2 in the blood due to energy demands of tissue such as contracting muscle pushes the reaction in the opposite direction. Carbon dioxide and water combine to form carbonic acid. Part of the carbonic acid dissociates into bicarbonate and $[H^+]$. This decreases blood pH and serves as a signal to central control centers to increase the rate of breathing. Increased breathing causes release of more carbon dioxide from the lung. As carbon dioxide is released into air, blood pH increases again toward neutral because the carbonic acid reactions proceed in the reverse order and breathing rate returns to baseline.

Summary

I hope that I have convinced you that water is more than just filler material around biologic molecules. Some of the material presented in this chapter is very hard to assimilate the first time through. But, these concepts will be used over and over again by your instructors in physiology to explain the complex interaction of the parts of the human body. It will be assumed that you already know this stuff very well.

Before you move to the next chapter, read the first chapter and this one a second time. You will get much more out of the second reading than the first. This is important, because material in the next chapter will be built upon these first two chapters. The sequence in which the ideas are presented was created to be a logical progression. Logical progressions form patterns, and patterns are easier for our minds to remember.

The major points you will need to remember from this chapter about water are:

- ❖ The polar electronic nature of water molecules allows them to form weak bonds with each other called hydrogen bonds that rapidly form, break, and re-form.
- ❖ Because the kinetic energy of the large number of molecules per volume of liquid water is high, small fluctuations of heat energy in the body cause little temperature change in the body's water.
- ❖ Only substances carrying an electrical charge or partial charge (dipoles) can interact with the dipole charges of water molecules and thereby dissolve in water.

- Ionic bonds such as those of crystalline NaCl or other salts are easily broken by water molecules.
- pH is a shorthand value used to describe the quantity of [H^+] in a water solution. [H^+] is a highly reactive ion when combined with water to form [H_3O^+] that is capable of randomly breaking chemical bonds.
- The primary system for maintaining neutral pH in human blood is the carbonic acid – bicarbonate mixture.

CHAPTER 3

HOW MOLECULES MINGLE AND RELOCATE – DIFFUSION, OSMOSIS, OSMOTIC PRESSURE, & HYDROSTATIC PRESSURE

Translation of the principles of chemistry for use in describing physiological function is not always readily apparent. Yet, the terms diffusion, osmotic pressure, and hydrostatic pressure are widely employed in explanation of how the work of living is handled by the systems of the body.

Physiology textbooks present a variety of explanations for what drives the transport of molecules within biologic systems. Some physiology textbook descriptions of osmotic pressure and hydrostatic pressure are totally conceptual, and some incorporate more or less of the equations of chemistry and physics that underlay these observable processes.

PHYSIOLOGY: CUSTOM-DESIGNED CHEMISTRY

In this chapter theories developed in previous chapters will be employed to explain the process of diffusion of molecules in solutions, the formation and composition of biologic membranes, osmosis, and osmotic and hydrostatic pressures in the human body.

It is important that you have read Chapter 1 and Chapter 2 before you begin this one. If you skipped those chapters, please go back and read them now.

What is diffusion?

Diffusion is the relocation of molecules from one part of a solution to another part. It is a relatively slow process, but it works well enough to redistribute molecules over short distances. Many important local effects of biologic molecules depend upon diffusion. Polar molecules, those where electrons are shared unequally between atomic nuclei, and electrically charged molecules diffuse readily through water. Nonpolar molecules such as oxygen (O_2) diffuse quickly through nonpolar materials like fats and oils.

Diffusion is sometimes described as a random walk of unbound molecules that occurs when two solutions merge. If the concentration, which is the number of dissolved molecules per unit volume, is greater for one of the solutions than other, there is said to be a *concentration gradient* between the solutions.

There are many ways of creating molecular concentration gradients in the human body. When a molecular concentration gradient exists, a greater number of dissolved molecules will diffuse from the area of high concentration to the area of low concentration than will diffuse in the opposite direction. You will sometimes see this referred to as molecules or *ions diffusing down their concentration gradient.*

Figure 10: Representation of the diffusion of molecules when two different solutions come in contact with each other. Ink in water. Photo © 2012 ADA photo

Diffusion, represented in the diagram above, in the presence of a concentration gradient continues until the dissolved molecules are evenly distributed and the gradient disappears. To visualize a concentration gradient, diffusion, and disappearance of the gradient, add a drop of food coloring to a glass of water and watch what happens.

Polar biologic molecules diffuse continuously within and between the body's fluid compartments. There are two major water compartments in the human body. The first includes all water contained within cells, the intracellular fluid compartment or cytosol. The second is the extracellular fluid compartment. It consists of all body water not in cells and is divided into two parts. The two parts are the interstitial fluid surrounding cells and the fluid contained in blood and lymph vessels.

In contrast to polar molecules, nonpolar molecules only diffuse in nonpolar solutions where molecules mingle closely together. Concentration gradient plays the same role in diffusion of nonpolar molecules in nonpolar environments as it does in diffusion of polar molecules in water solution.

What is a semi-permeable membrane?

In order to have a discussion about *osmosis, which is a way to move molecules of water from one solution to another,* we need to first discuss biologic membranes. The aversion of nonpolar molecules to diffusion in water allows biologic systems to create water-filled compartments separated from each other by barriers composed of nonpolar, hydrophobic, molecules. Such barriers between body compartments made up of specialized hydrophobic molecules are called *biologic membranes*.

The major components of biologic membranes include various types of hydrophobic molecules. Other minor components include a small number of mixed-quality molecules that have a hydrophobic exterior and a hydrophilic core. Mixed-quality molecules may span from one side of the membrane to the other. Part of their exterior is hydrophobic and it permits them to diffuse well in the hydrophobic membrane. Their hydrophilic core is sometimes shaped like a tunnel passing through the membrane. The tunnel, or pore, is large enough for either a water molecule or another hydrophilic molecule to pass.

The hollow structure represented in the model membrane on the next page depicts a pore used by hydrophilic molecules to pass through a hydrophobic cell membrane. Because individual pores are discriminatory about what they will let pass, biologic membranes are said to be *semi-permeable membranes*.

For a membrane to be permeable to several classes of hydrophilic molecules, it must have distinct pores for each class. Membrane pores are uniquely shaped to allow only one type of molecule to pass through. For example, a sodium ion cannot pass through a pore designed for a potassium ion. Cell membranes have one set of pores for sodium and another set of pores for potassium. This allows passage of sodium through the membrane to be regulated independent of the passage of potassium.

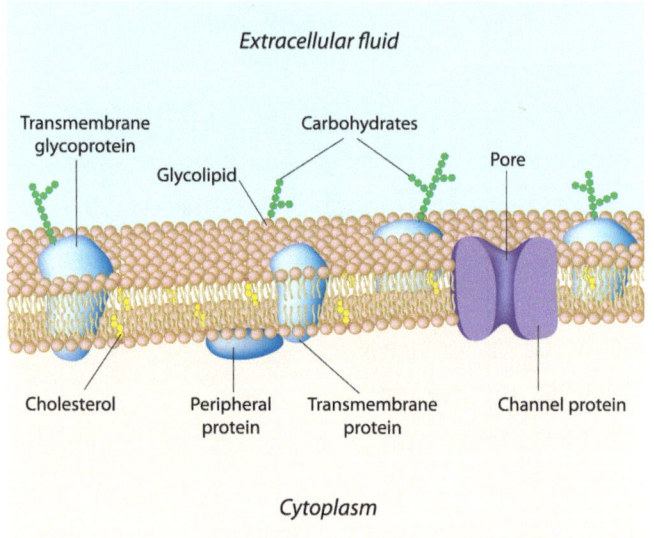

Figure 11: A model of a cell membrane showing some common components. Photo © 2012 Alila Medical Images

An example of the actual protein structure of an ion pore for potassium is pictured on the next page. Proteins are made up of 20 different small molecules called amino acids some of which are hydrophilic and some hydrophobic. Individual proteins range in size from about 100 amino acids to several thousand amino acids.

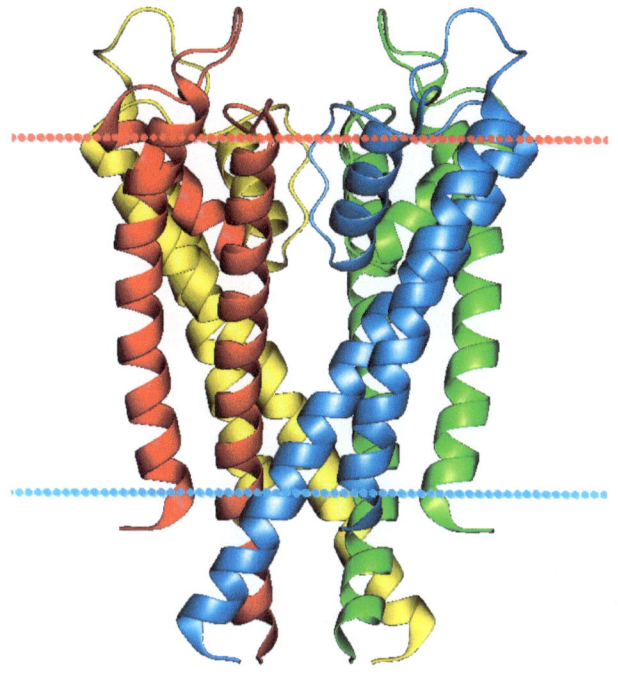

Figure 12: Molecular structure of a simple potassium ion pore found in bacteria. Image data is from OPM database. Photo © 2007 Andrei Lomize

The protein molecule creating this potassium pore of bacteria is represented as a ribbon model displaying the shape of the molecule's central carbon chain. Each helical turn of the carbon chain is made up of 4 amino acids. The parallel lines represent the boundaries of the nonpolar plasma membrane molecules. The portion of the ribbon between the lines is composed of hydrophobic amino acids where it contacts the membrane's molecules. The portion of the model extending beyond the membrane into the cell's cytoplasm and into the extracellular fluid is made up of hydrophilic amino acids. The opening in the center is also composed of hydrophilic amino acids and it permits passage of only K^+.

What is osmosis?

Osmosis is the flow of water molecules through a water pore in a semi-permeable membrane. If the membrane studied only has pores that accommodate water, then the amount of water that flows through those pores will depend on the energy level of the water compartments on either side of the membrane.

Initially in the experiment diagramed on the next page, on each side of the semi-permeable membrane there are two water solutions of equal volume but unequal number of dissolved molecules. In other words there is a molecular concentration gradient between the two sides of the membrane.

If the membrane were permeable to the dissolved molecules, the molecules would move through it by diffusion and eventually eliminate the concentration gradient. But the membrane is not permeable to the dissolved molecules. It is only permeable to the water.

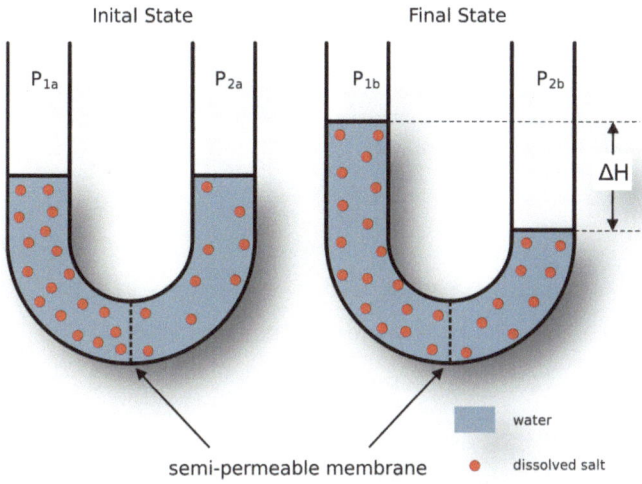

Figure 13: A diagram of a simple chemistry experiment designed to demonstrate osmosis. Photo © 2011 Hans Hillewaert

Let's take a minute to remember something we learned in Chapter 1. There we discovered that molecules bonded to each other are at a lower energy level than unbonded molecules. That concept can be extended to explain the result of a laboratory osmosis experiment.

Hydrogen bonding activity occurs between water and dissolved molecules. The greater the number of dissolved molecules, the greater the degree to which the water molecules are in a bonded state. The greater the number of hydrogen bonded water molecules the lower the kinetic energy of the water compartment. Because of hydrogen bonding activity, the water molecules on the side of the membrane with fewer dissolved molecules possess more kinetic energy than the water molecules on the other side of the membrane.

Osmotic water flow is always from the side of the membrane with fewer dissolved molecules (high energy water molecules) to the side with more dissolved molecules (low energy water molecules). It is the difference in kinetic energy of water molecules on either side of the membrane that dictates the direction of flow of water. When enough water has crossed the membrane so that kinetic energy of the water molecules is equivalent on both sides of the membrane, water flow will cease.

What is osmotic pressure?

Osmotic pressure is a way to add a numerical value to the osmotic flow of water. Osmotic pressure is often defined to be the amount of pressure that must be added to prevent osmosis – the water flow across a membrane separating two solutions with differing concentration of dissolved molecules and different energy levels. The greater the initial difference in solute concentration on the two sides of the membrane, the greater the number of water molecules that will flow across before kinetic energy equalizes.

HOW MOLECULES MINGLE

If the driving force of osmosis is thought of as energy differences of water molecules on two sides of a semi-permeable membrane, how does adding pressure to the more highly concentrated side of the membrane stop the water flow? To understand this, first we need to look at the equation for pressure exerted by a static fluid. It is:

$$P_{\text{static fluid}} = \rho g h$$

P = pressure; ρ = m/V = fluid density; g = acceleration of gravity; h = depth of fluid

In the osmosis experiment pictured in Figure 13 above, the height of the columns of water on both sides of the membrane changed from the initial to final state. Movement of water across the membrane with osmosis increased the depth of one of the columns and decreased the depth of the other. In the equation above 'h' changed. Therefore, influx of water into the more concentrated solution, according to the equation for pressure in a static fluid, created a fluid pressure difference on the two sides of the membrane.

We all know that water can be moved through an opening, such as a pipe, by adding pressure to the water. This is because pressure exerted on one part of a fluid volume distributes throughout the entire volume. Water comes into your home because there is a source of water under pressure that is connected to your house. For our purposes here, a pore in a membrane can be thought of as a tiny pipe.

At the end of the osmosis experiment above, theoretically it should be possible to move water molecules back through the membrane by adding external pressure to the column with a greater depth of fluid. Also, in theory the amount of pressure needed to restore the experimental columns back to their initial height should be equal to the pressure difference between the columns created by the osmotic flow of water.

This is easy to test experimentally. You would first calculate pressure difference between the two columns after osmosis occurred using the equation above. The next step would be to set up the initial conditions again, but this time you would also externally add the calculated amount of pressure to the more concentrated volume of water.

When an experiment is set up in this fashion osmotic flow of water does not occur. The added external force precisely offsets the pressure difference of the two columns of water created by flow of water across the membrane. The magnitude of the required pressure to stop water flow is called the osmotic pressure of the more concentrated solution.

It is common practice to state osmotic pressure in units of millimeters (mm) of Hg and centimeters (cm) of

H_2O. To make comparisons of pressures in fluid systems, all pressures are converted to units equal to the effect of gravity on static fluid column heights such as mm of Hg and cm of H_2O. In physiology textbooks, osmotic pressure is usually stated as a number of mm of Hg.

Another way of thinking of osmotic pressure is that it is an indication of the force with which pure water moves across a biologic membrane as a result of the solute concentration of a body fluid. In physiology, when an osmotic pressure value is reported for a particular biologic fluid compartment, that value represents the force with which pure water would theoretically move into that compartment.

Physiology uses the principles of osmosis and osmotic pressure to predict movement of water from one fluid compartment in the body to another. If a body compartment is stated to have a high osmotic pressure, that compartment has a high potential for drawing water into it.

In practice, it is the comparison of the osmotic pressure of two adjoining biologic fluid compartments that dictates flow of water. For example, if compartment A has an osmotic pressure of 25mmHg and the adjoining compartment B has an osmotic pressure of 20mmHg; water will flow from compartment B that has the lower osmotic pressure into compartment A that has the higher osmotic pressure. The osmotic pressure driving the flow of water in this case is 5mmHg – the magnitude of the difference between the osmotic pressure of the two compartments.

A specific example of the role of played by osmotic pressure in the body is the large loss of water from the extracellular fluid compartment, dehydration, due to excessive exercise in hot weather. During dehydration the fluid bathing the body's cells becomes more concentrated as water is lost to the environment and it develops a higher than normal osmotic pressure. Water then flows from the cells into the overly concentrated extracellular fluid. With restoration of body water by drinking, osmotic pressure of the extracellular fluid lowers to normal and water moves back into cells until osmotic pressure of the two compartments equalizes.

How does hydrostatic pressure differ from osmotic pressure?

Hydrostatic pressure is another concept used to describe the movement of water across biologic membranes. Hydrostatic pressure develops in a volume of water, or water solution, when an outside pressure is applied to that volume. But, it is different than osmotic pressure. Osmotic pressure is a measure of a water solution's ability to draw additional water into it because of its dissolved molecules. *Hydrostatic pressure of a water solution depends on application of outside pressure, such as gravity, and it is not dependent upon the number of dissolved molecules.*

It is a challenge to convert the physics definition of hydrostatic pressure for use in physiology. In physics, hydrostatic pressure is defined as the pressure exerted by a fluid at rest due to the force of gravity – as in the equation in the above section on osmotic pressure.

Water towers, for example, stand high above ground and gravity exerts a force that can drive that water toward the ground. The energy, or hydrostatic pressure, with which the water will leave the tower when a valve opens in its bottom is proportional to the depth (and therefore weight) of the water.

In physiology, hydrostatic pressure is defined as the pressure in the blood circulatory system that is exerted by a volume of blood confined to a blood vessel. As in the case of the water tower, blood pressure results from the weight of the liquid in the system. The driving force in biology, however, is not gravity and the fluid – blood – is not at rest but flowing.

Instead of gravity being the driving force, pressure is exerted on the blood volume in blood vessels by the heart pump. Hydrostatic pressure created by the heart causes blood to flow into large exit vessels that must expand to accommodate the local volume. Recoil pressure of elastic components in the vessel walls force blood to flow toward vessels that repeatedly divide into two somewhat smaller vessels. Hydrostatic pressure in the vessels distant from the heart drops, because with each division total blood vessel volume increases and the hydrostatic pressure is distributed over a larger area.

By the time blood reaches the smallest vessels called capillaries, where nutrient exchange occurs between blood and fluid bathing cells, hydrostatic pressure is comparatively low. It must be low, because these very thin walled vessels necessary for diffusion of nutrients cannot survive a great deal of force.

However, to say capillary hydrostatic pressure is low compared to hydrostatic pressure in vessels nearer the heart is not to say it is too small a force to drive water out of capillaries into tissue. In fact, it is capillary hydrostatic pressure balanced against blood osmotic pressure that permits exchange of nutrients for cell waste molecules in tissue.

As long as hydrostatic pressure, the force that pushes water through the capillary membrane, is sufficient to more than balance osmotic pressure of the blood that draws water into the capillary, blood water will move into tissue carrying nutrients with it.

As water is lost from the capillary to the fluid that surrounds cells, dissolved molecules in blood become more concentrated. This increases the osmotic pressure of the blood. Eventually, hydrostatic pressure pushing water out of the capillary will no longer be strong enough to oppose blood osmotic pressure. When blood osmotic pressure becomes the greater force, water will flow from the interstitial fluid into the capillary. Interstitial water entering the capillary carries cell waste with it. Cell waste collected by the blood at the capillaries is removed later by the kidney.

Summary

To summarize there are several key concepts to remember from this chapter.

- ❖ When a concentration gradient exists for a molecule in solution, that molecule will move away from the area of its high concentration – this is described as a molecule diffusing down its concentration gradient.
- ❖ For a polar molecule to cross a biologic membrane, the membrane must have a pore specifically designed for that molecule. Membrane pores are very discriminatory causing membranes to be described as semi-permeable.
- ❖ Osmosis is the flow of water molecules through a water pore in a semi-permeable membrane.
- ❖ Osmotic flow of water is from the side of the membrane with higher energy water mole-

cules (less dissolved molecules) to the side with lower energy water molecules (more dissolved molecules).

❖ Osmotic pressure is a numerical value that describes the osmotic flow of water. In physiology, the determinant factor controlling flow of water across a membrane is the difference in osmotic pressure of adjoining compartments.

❖ In physiology, hydrostatic pressure is defined as the pressure in the blood circulatory system that is exerted by a volume of blood confined to a blood vessel.

❖ Whichever is greater, blood hydrostatic pressure or blood osmotic pressure, will dictate flow of water into and out of blood capillaries.

CHAPTER 4

PHYSIOLOGY'S INTERFACE WITH EARTH'S ATMOSPHERE – EXPLORING GAS LAWS

This chapter is devoted to a segment of chemistry called the gas laws. The gas laws usually appear for the first time when lung physiology is introduced. Having a good understanding of the gas laws before beginning your study of lung physiology will be advantageous. It will provide you a considerable head start in learning the processes that the human body uses to exchange oxygen (O_2) for carbon dioxide (CO_2).

In previous chapters, physiological processes were discussed in terms of water's liquid environment. But, water's liquid chemistry is only part of the story. To fully understand physiologic function, we also must explore rules that govern Earth's atmospheric gas molecules – water vapor, oxygen, carbon dioxide, and nitrogen.

PHYSIOLOGY: CUSTOM-DESIGNED CHEMISTRY

The first section of this chapter will provide a brief description of the role of carbon dioxide and water in capture of the sun's energy and the role of oxygen in the use of that energy. This section is included because it is important to remember why it is that the body must gather oxygen in from Earth's atmosphere and return carbon dioxide to it.

Critical reactions within the human body to extract energy from food rely upon having oxygen available on site. Diffusion of free molecules is only effective over short distances as was discussed in Chapter 3. So, moving oxygen gas molecules from the atmosphere to sites deep within the body depends upon creative use of nature's laws of gas behavior.

How are gasses of Earth's atmosphere used to store and retrieve the sun's energy for physiology?

The overall process for making sugar by using the sun's energy (photons of light), water, and carbon dioxide is called *photosynthesis*.

Photosynthesis consists of a series of light-dependent and light-independent reactions in plants. The Calvin Cycle, the light-independent reactions, uses the energy of the molecules created in the light-dependent reactions to transform carbon dioxide gas and water into sugar.

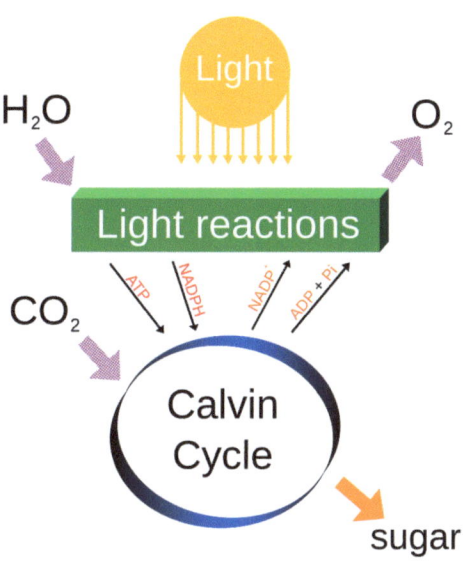

Figure 14: Simplified diagram of the components of photosynthesis and respiration. Photo © 2008 Daniel Mayer

The energy transfer steps are complex but the overall photosynthesis reaction can be written:

$$6\ CO_2 + 6\ H_2O + \text{photons} \longrightarrow C_6H_{12}O_6\ (\text{sugar}) + 6\ O_2$$

Plant sugar is eaten by humans and other animals. It is the primary source of energy to drive physiological processes. In chapter 1 we talked about energy released when bonds holding together the atoms of sugar molecules are broken in a controlled multistep process. The energy release process is called *respiration* by biochemists. The overall reaction is:

$C_6H_{12}O_6$ (sugar) + 6 O_2 —› 6 CO_2 + 6 H_2O + 2880 kJ/mole

Energy is released to the system when sugar is converted to carbon dioxide and water, because sugar's constituent atoms of carbon, oxygen, and hydrogen retain greater kinetic energy when bound in a sugar molecule than they do bound as carbon dioxide and water. That means that the formation of carbon dioxide and water releases more energy than it takes to break the weak (high kinetic energy) bonds of sugar.

The net effect of plant photosynthesis followed by human and animal respiration is conversion of the sun's light energy into a form that is useful for powering physiological processes. Two atmospheric gasses are vital to this set of molecular conversions – oxygen and carbon dioxide. Therefore, it is important to define the laws that govern availability of these two gasses.

What are the basic assumptions of the gas laws?

GAS LAWS

The kinetic energy theory of atoms and molecules is most clearly stated in the assumptions that describe how an ideal gas behaves. Good theories based upon realistic assumptions predict real life events reasonably well. The ideal gas laws provide physiologists with an excellent starting point for explaining delivery of oxygen to cells deep in the body.

The characteristics of an ideal gas are defined to be:

- Molecules are very far apart relative to their size
- The number of molecules is so large that statistical treatments are valid
- Molecules are in constant random motion
- Molecules collide frequently with each other and with the wall of any container they may be within
- Collisions between molecules are elastic – they hit each other and bounce off
- Some energy is transferred from one molecule to another but total energy in the system is constant
- Molecules are considered to be spherical in shape and behave as if they all have the same mass
- Time during collision of molecules with a container's wall is negligible compared to the time between successive collisions

- There are no forces of attraction or repulsion between molecules – unlike water molecules in the liquid form
- Average kinetic energy of molecules depends upon temperature – warmer equals faster moving

Oxygen, nitrogen, and carbon dioxide and even water vapor at normal air and body temperature behave like ideal gasses. The polar forces that cause water molecules to hydrogen bond in a liquid phase are generally overcome by the distance between water molecules in water vapor.

What is gas pressure?

Biologic systems depend on the force created by the kinetic energy of gas molecules to move oxygen from the atmosphere into the body. When gas is within a container, the random collisions of gas molecules with the wall of the container create a force that can be measured. The measured force divided by the area of collision is defined to be the pressure of the gas.

Earth's gas atmosphere is the human body's source of oxygen. Earth's atmosphere exerts a gas pressure on things it contacts due to the kinetic energy of its component gasses. The four gasses that make up earth's atmosphere each act much like an *ideal gas* at normal temperatures. The Earth's atmosphere is composed of about 78% nitrogen, 21% oxygen, a varying percentage of water vapor, and a variety of other trace gasses including about 0.039% carbon dioxide.

The amount of pressure generated by a gas mixture is influenced by the number and kinetic energy of the component gas molecules in a given volume. Because gasses move faster at higher temperature, temperature can be used as a relative quantifier of gas energy. That is, the higher the temperature of gas in a container, the higher the pressure that the gas will exert on the container.

There are three gas laws that describe the relationship between gas pressure, gas volume, and temperature of *ideal gasses*. The laws are *Boyle's Law, Charles's Law, and Gay-Lussac's Law*. These three gas laws can be mathematically combined into a *Combined Gas Law* that can be written:

$$P_1V_1/T_1 = P_2V_2/T_2$$

The Combined Gas Law is useful for describing gas exchange between the human body and the atmosphere at the lung. All three elements in this equation vary as atmospheric air moves in and out of human lungs. The temperature of air in the lung will generally be different than atmospheric air temperature. The volume occupied by air in the lung changes with contraction and relaxation of the muscles of the rib cage and the diaphragm. With enlargement of the lung compartment, pressure of the lung-enclosed gas will decrease as volume increases. In contrast, when muscle relaxation decreases the size of the lung compartment, pressure of lung-enclosed gas will increase.

Why does gas flow from place to place?

The high kinetic energy of gas particles in general causes gas to expand into any available container. If there are two containers with different pressures, *the net flow of gas particles will be from the container with higher gas pressure to that with lower gas pressure.*

For example, when lung volume expands with enlargement of the thoracic compartment, pressure inside the lung drops below that of atmospheric gas pressure and atmospheric gas expands (flows) into the lung until pressures equalize. When the rib cage and diaphragm return to their relaxed state, the lung's smaller volume drives the lung's gas pressure above atmospheric gas pressure and air flows out of the lung.

What does 'partial pressure' of a gas mean?

When there is a mixture of gasses, as in earth's atmosphere, each gas contributes to the total pressure based upon its percentage of the blend. The partial pressure of a gas in a mixture is defined by *Dalton's Law of Partial Pressures* (circa 1801) to be the pressure which the gas would have had if it had been alone occupying the volume.

It is because of the great spaces between the randomly moving ideal gas molecules that each type of gas can be treated as if the others were not present. The pressure created by the kinetic energy of an individual gas's molecules on a surface is not influenced by the other molecules present, because they are too far apart to interact with each other.

Dalton's Law of Partial Pressures allows physiologists to treat oxygen gas and carbon dioxide gas as separate entities. The flow of oxygen into the lung and carbon dioxide out of the lung can be predicted based upon a comparison of the partial pressure of oxygen and carbon dioxide in the lung with the partial pressure of oxygen and carbon dioxide in the atmosphere.

What influences the magnitude of atmospheric partial pressures?

The sum of addition of the partial pressures of the component gasses of a mixture of gasses must equal the total pressure of the mixture. That is, the sum of the parts must equal the whole. Therefore, the sum of the

PHYSIOLOGY: CUSTOM-DESIGNED CHEMISTRY

partial pressure of all of the gasses in the atmosphere must equal atmospheric pressure.

By actual measurement, at sea level atmospheric pressure is equal to the force needed to drive a column of mercury to a height of 760 mm. That means the sum of the partial pressure of water vapor, oxygen, carbon dioxide, nitrogen, and other atmospheric gasses must equal 760 mmHg at sea level.

However, Earth's atmosphere does not have a uniform pressure overall. It varies from about 760 mmHg at sea level to about 440 mmHg at the top of Mt. Whitney (14,505 feet above sea level) in California. Atmospheric pressure decreases at altitudes high above sea level, because the number of gas molecules per unit volume is smaller at higher altitudes.

It is important to remember that the percentage distribution of component gasses does not change with altitude or with temperature. For example, oxygen is 21% of atmospheric gas both at sea level and at high altitude. This means oxygen's partial pressure at sea level is equal to about 159.6 mmHg compared to about 92.4 mmHg on Mt. Whitney. That is because 21% of 760 mmHg equals 159.6 mmHg and 21% of 440 mmHg equals 92.4 mmHg.

On Mt. Whitney it is harder to get adequate oxygen flow into the lungs during exercise, because oxygen partial pressure in the lung must fall to a lower than normal level to get a normal amount of atmospheric oxygen to flow into the lung. Remember that it is the difference in partial pressure between the atmosphere and the lung that causes gas flow.

How does nonpolar oxygen move through even short distances in water?

The lung like the rest of the human body is a water environment. There are large air spaces in the lung, but the cells, like other cells in the body, are bathed in interstitial fluid. Oxygen and carbon dioxide in the lung must cross this water barrier on their way to and from the blood capillaries. Once at the cell membranes both of these molecules diffuse rapidly through the nonpolar membrane layers.

Gas pressure differences, responsible for net flow of particles from one container to another with expansion of a gas, also cause movement of gas molecules into and out of solutions. *Henry's Law* of gasses states that the concentration of dissolved gas in a solution is directly proportional to the partial pressure of the gas above the solution.

There are some restrictions on Henry's Law. It is only true if the molecules of liquid and gas are in a state of equilibrium. Equilibrium exists when gas molecules entering the liquid phase equal the number of molecules exiting. Henry's law also does not work for gasses at high pressures, but it does work for gas in the human lung, because the human lung is a very low pressure system.

Oxygen is a nonpolar molecule and it does not diffuse in water in the normal sense of a random walk as described in Chapter 3. Rather oxygen slips into pockets that exist in the loose hydrogen-bonded network of water molecules without forcing them apart. The overall size of the pockets between water molecules suits the shape and size of oxygen molecules. The oxygen is then caged by water molecules, which weakly pin it in place until the dynamic interaction of the water molecules cause the pocket to disappear. Then the partial pressure of oxygen over the water pushes oxygen into another available water pocket.

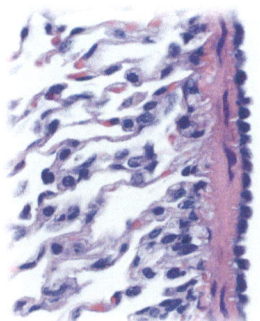

Figure 15: A photo-micrograph of a histologic section of lung tissue showing alveoli (thin blue cells surrounding open spaces) beneath the lung surface. Photo © 2012 Vetpathologist

The small pink areas attached to the thin walls of the alveoli are lung capillaries. They are very small and tightly attached to the wall of alveoli.

Once oxygen reaches the lung cell and blood capillary membranes it diffuses rapidly through the nonpolar environment of membranes into blood. In the blood capillary, oxygen moves into the red blood cells by diffusion and binds to hemoglobin molecules. It is carried by the red blood cells to parts of the body where the partial pressure of oxygen is lower. Encountering the lower tissue partial pressure allows oxygen to diffuse out of the red blood cells. Oxygen then enters into the interstitial water reversing the process at the lung until it reaches membranes of the tissue's cells where it diffuses into the cells.

The transfer of carbon dioxide from interior body cells into blood and then to lung is a more complex process than the delivery of oxygen to those cells. Solubility of carbon dioxide in water is even less than that of oxygen. It does not fit as well as oxygen in water pockets.

Efficient movement of carbon dioxide into blood and delivery to the lung uses a set of reactions that involves the carbonic acid-bicarbonate buffer system described in Chapter 3. However, once carbon dioxide's partial gas pressure rises in the lung, it flows out to the atmosphere obeying the same gas partial pressure laws that regulate oxygen's movement into the lung.

Earth's atmosphere is only about 0.039% carbon dioxide which is a partial pressure of about 0.3 mmHg. The very low partial pressure of carbon dioxide in air assures that carbon dioxide reaching the lung from deep tissues will flow out into the atmosphere. Lung alveolar air usually has a carbon dioxide partial pressure of about 40 mmHg.

Summary

To summarize there are several key concepts to remember from this chapter.

- ❖ The net effect of plant photosynthesis followed by animal respiration is conversion of the sun's light energy into a form that is useful for powering physiological processes.
- ❖ The gas in the Earth's atmosphere displays characteristics of an ideal gas.
- ❖ The laws describing the relationship between gas pressure, gas volume, and temperature of ideal gasses, *Boyles Law*, *Charles's Law*, and *Gay-Lussacs's Law*, can be mathematically combined into the *Combined Gas Law* written: $P_1V_1/T_1 = P_2V_2/T_2$

GAS LAWS

- The high kinetic energy of gas particles in general causes gas to expand into any available container. The net flow of gas particles will be from a container with higher gas pressure to that with lower gas pressure.
- *Dalton's Law of Partial Pressures* states that the partial pressure of a gas in a mixture of gasses is defined to be the pressure which the gas would have had if it had been alone occupying the volume.
- The sum of the partial pressure of all of the gasses in Earth's atmosphere must equal atmospheric pressure.
- Because Earth's atmospheric pressure changes with altitude, the partial pressure of each gas in the mixture also changes with altitude.
- *Henry's Law* of gasses states that the concentration of dissolved gas in a solution is directly proportional to the partial pressure of the gas above the solution.
- Both oxygen and carbon dioxide are nonpolar molecules that are driven into pockets between water molecules by their partial pressures. But, the shape of oxygen molecules fits better in water pockets than the shape of carbon dioxide molecules.

PHYSIOLOGY: CUSTOM-DESIGNED CHEMISTRY

❖ Movement of carbon dioxide from cells where it is made to the lungs involves a complex set of reactions employing the carbonic acid-bicarbonate buffer system described in Chapter 2.

CHAPTER 5

FLUID COMPARTMENTS: THE PLATFORM FOR LONG DISTANCE COMMUNICATION

One way that physiology deals with the problem of its overwhelming complexity is to create specialized compartments. One of the efficiencies gained by doing this is that identical biological molecules can have many different jobs depending upon the compartment in which they are located. To a great extent physiology courses are dedicated to learning how various functional compartments maintain their integrity and communicate with each other.

Theories of physiology use the chemistry principles presented in previous chapters to accurately describe cross talk between the body's compartments. As a first step in understanding physiology, you must discover the anatomical boundaries of each compartment. If the boundary is a cell membrane, then ask what can pass through that membrane. If the boundary is a blood vessel, then ask what can pass through the vessel wall. Next you want to ask about the characteristics of the fluid environment on each side of that boundary.

Now you are ready to predict which molecules on each side of the boundary will pass through based upon the principles learned in Chapters 1 through 4. What kind of pores does a particular cell membrane have for movement of water-soluble molecules? Are the pores always open? What causes the pores to open and close? Does regulation of the pores in the membrane require energy or are they passive? What pressures cause molecules to move across the membrane? In which direction do molecules pass? How does diffusion of molecules across a membrane facilitate communication over long distances?

By the end of this chapter you should understand why you would want to ask such questions in the first place. Here you will discover how the fluid compartments of the human body set the stage for long distance communication between body parts.

What are the fluid compartments of the human body?

At first exposure, the composition and dynamic nature of the body's three water compartments appears simplistic and boring. When memorizing lists of ions and other molecules present in each fluid compartment, it is difficult to imagine their relevance to moving a finger, beating of a heart, or being able to think. Yet all of such physiological activity actually does depend on the concentration of particular molecules in each of the fluid compartments. And, behavior of those molecules is dictated by the chemistry described in chapters 1-4 of this book.

There are three fluid compartments in the body to be considered. The first is the *intracellular fluid* often referred to as cytosol. The second is the fluid outside and between the cells called the *interstitial fluid*. And last there is the vascular fluid, *blood or lymph*, contained in long tubes running through the interstitial fluid.

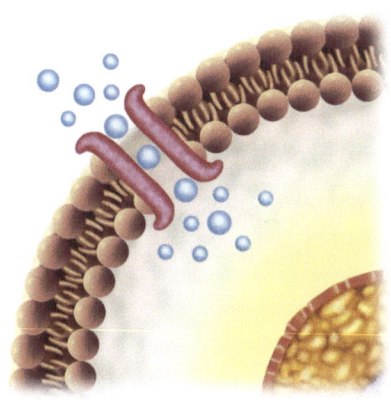

Figure 16: Model of a cell membrane with water molecules moving through an open pore. Photo © 2012 Alex Luengo

Membranes separate these compartments but water circulates between them in response to osmotic pressure differences. Each of these compartments has a very distinct character. The unique nature of each depends upon the concentration of specific dissolved molecules and ions. A large amount of chemical energy and anatomic creativity is invested in stabilization of fluid compartment components.

An example of a boundary between fluid compartments – a capillary wall separating blood and interstitial fluid – was presented in Chapter 3. There the effect of a balance between hydrostatic and osmotic pressure was discussed. In contrast to normal, if blood and interstitial fluid possessed equal osmotic pressure, water driven out of the capillary by hydrostatic pressure would not be returned to the capillary at its venous end. The closed circulatory system would then collapse, because blood volume would rapidly shrink.

To understand the relationship between the three fluid compartments, it may help you to think of the cells of the body as floating in a protein free watery solution called *interstitial fluid*. Two types of vessels run through the interstitial fluid surrounding cells – blood containing vessels and lymph containing vessels. Blood vessels bring nutrients and oxygen to the interstitial fluid for diffusion into cells. Floating cells deliver carbon dioxide and waste molecules to the interstitial fluid. Blood takes up carbon dioxide and delivers it to the lungs for release from the body. Blood also carries away water soluble cell waste that enters with water at the ends of the capillaries. The lymphatic vessels remove remaining cellular waste material and any material that leaks from the blood vessels.

Blood is made up of a water solution called plasma that contains ingested nutrients, a number of very large biologic molecules called proteins, ions, red blood cells, white blood cells, and cellular waste material. Lymph is interstitial fluid collected from around the cells by lymph capillaries. Lymph has a higher osmotic pressure than interstitial fluid. As lymph circulates through the body, it also picks up fat from the intestine and white blood cells from the lymph nodes where it mixes with blood. Lymph vessels are also very efficient at recovering from interstitial fluid any proteins that leak out of blood vessels or cells. It is the lymphatic system that assures that the osmotic pressure of interstitial fluid remains very low.

The lymphatic system is an open system. There is no pump like the heart to drive lymph through the vessels. Lymph movement is due to contraction and relaxation of smooth muscle in the wall of the vessels and contraction of adjacent skeletal muscle. Valves in lymph vessels prevent backflow. Eventually lymph vessels join with the large veins bringing blood to the heart and lymph is added to the blood. Interstitial fluid removed by lymph vessels is balanced by water and soluble material entering the interstitial fluid from blood capillaries. The final decision about which small molecules and ions the body should keep in the blood and which it should eliminate is made as the blood passes through the kidney.

What is an electrochemical gradient?

An *electrical gradient* across a membrane separating two fluid compartments exists when one surface of the membrane has a positive charge and the other surface has a negative charge. By definition a voltage is a difference in electric potential between two points. Sensitive volt meters can measure voltage differences across cell membranes.

In physiology the voltage difference between the inside and outside of a cell membrane is called a *transmembrane potential*. By convention cell membrane voltage is always read as inside electrical potential relative to outside. For example, if a transmembrane voltage is minus 70mV, the electrical potential of the inside surface of the cell membrane is 70mV less than that of the outside surface of the cell membrane. Transmembrane potentials are created by the diversity of membrane pores and the concentration of electrically charged molecules and ions in the fluid compartments on either side of the membrane.

It is common practice when discussing electrochemical gradients to refer to specific ion pores as ion channels. So, in the remainder of this chapter the term 'channel' will be equal in meaning to the term 'pore' in previous chapters.

A *chemical gradient* exists when the concentration of a particular molecule differs in the fluid compartments on either side of a membrane. Remember from Chapter 3 that molecules diffuse from an area where their number per volume is high to an area where their number per volume is low. In physiology *diffusion down a chemical gradient* often occurs across open channels in cell membranes.

The term *electrochemical gradient* was devised to describe the combination of chemical and electrical forces that drive electrically charged potassium (K^+), sodium (Na^+), chloride (Cl^-) and calcium (Ca^{++}) through channels in cell membranes. If an open membrane channel is available for a charged ion such as K^+, the amount of K^+ that moves through the channel will depend upon two factors. The first factor is the chemical gradient between the fluids on either side of the membrane. The second factor is the electrical charge distribution on the membrane. K^+ will be attracted by a negatively charged membrane surface and be repelled by a positively charged membrane surface.

Chemical ion concentration gradients for potassium, sodium, and chloride exist across membranes that separate intracellular fluid from interstitial fluid. The fluid inside cells has a very high concentration of K^+ and of large negatively-charged molecules, but it has a very low concentration of Na^+ and Cl^-. The level of Ca^{++} inside cells varies by cell type, and most of it is constrained – it is only free to move intermittently. Outside cells Ca^{++} is very low in interstitial fluid. In contrast to intracellular fluid, the interstitial fluid in which cells float is very high in Na^+ and in Cl^- and very low in K^+. Large organic molecules are absent from interstitial fluid.

When open channels in the cell membrane are available for K^+ to pass, K^+ will diffuse out of the cell down its concentration gradient. When channels are open for Na^+ or Cl^-, these two ions will diffuse into the cell down their concentration gradients. The charge on the membrane, however, acts as a brake to diffusion. An ion will only flow through its channel until the concentration difference favoring diffusion is opposed by the charge on the membrane surrounding the channel. Using K^+ again as an example, K^+ moves out of a cell through its open channel until the negative charge on the inside of the membrane is great enough to hold K^+ in the cell in spite of a continuing diffusion gradient favoring flow of K^+ out of the cell.

FLUID COMPARTMENTS & COMMUNICATION

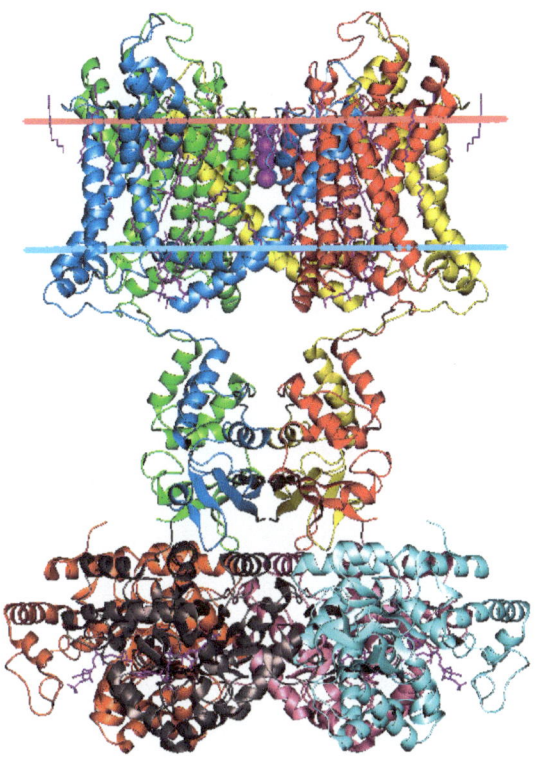

Figure 17: The experimentally determined complex structure of a potassium channel found in eukaryotic cells. Image data is from OPM Database. Photo © 2007 Andrei Lomize

What causes a transmembrane potential?

Pictured above is a model of a potassium channel's protein structure as it spans an animal cell plasma membrane. Calculated membrane boundaries are indicated by the parallel lines.

A transmembrane potential is caused by the selective permeability of cell membranes for ions. Cell membranes are leaky to a small extent. That is, there are always a few channels for ions like K^+, Na^+, and Cl^- that are open. However the quantity of leaky channels for K^+ is by far the greatest. That is, cell membranes in general are much more permeable to K^+ than to Na^+ or Cl^-. And, it is diffusion of K^+ out of the cell that contributes most to establishment of the transmembrane potential.

Through a cell's leaky channels, more K^+ atoms diffuse out of the cell down potassium's concentration gradient, than Na^+ atoms move into the cell down sodium's concentration gradient. That is, positive charges leaving the cell are greater in number than positive charges moving into the cell. The excess positive charge developed on the outside of the membrane as K^+ leaves the cell and lines up along the membrane surface electrically drawn by excess negative charge created on the inside of the membrane by loss of K^+ ions.

Transmembrane potentials caused by leaky or *'passive'* channels are said to be *equilibrium cell membrane potentials*. It takes very few charged ions passing through membrane channels to create a transmembrane potential. K^+ ion flow is limited by two factors. The first factor is the increasing positive charge on the outside of the membrane that repels K^+. The second factor is the buildup of negative charge on the inside of the membrane that attracts K^+. The transmembrane potential recorded when the electrical and chemical forces on K^+ balance, and further net flow of K+ through its membrane channel ceases, is called the *equilibrium potential for potassium*.

Over time passive membrane ion channels would eliminate concentration gradients between intracellular and interstitial fluid compartments if left alone.

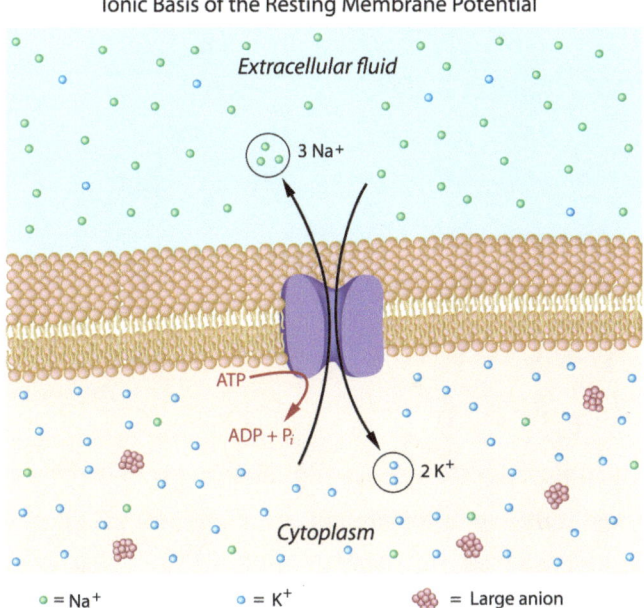

Figure 18: Energy requiring pumps maintain membrane potential. Photo © 2012 Alila Medical Images

To solve this potential problem, cell membranes also have energy requiring ion pumps that work constantly to maintain intracellular and interstitial fluid ion concentrations at their proper level. The energy for these pumps comes from cell respiration (Chapter 4).

The combined effect of passive ion channel equilibrium potentials and ion pump activity creates a cell's *resting potential*. Different types of cells have different resting potentials, because of diverse patterns of passive ion channels and energy requiring ion pumps in their plasma membrane. Examples of resting membrane potentials reported in literature include:

- o Skeletal muscle cells: -95mV
- o Smooth muscle: -60mV
- o Neurons: -60 to -90Mv

How are transmembrane potentials manipulated to create long distance communication within the body?

It is possible to send a signal along a cell membrane by sequentially altering the transmembrane potential. Manipulation of the transmembrane potential requires the presence of a variety of ion channels that can be opened and closed in an orderly fashion. For example, opening more Na^+ channels would reset a transmembrane potential. It would become less negative/more positive than the cell's resting membrane potential, because the outflow of K^+ and the inflow of Na^+ would be more evenly balanced. If the change in transmembrane potential away from the resting value can be used to create more open channels a little further away on the membrane, then there is the beginning of an electrical signaling system along the membrane.

Beside passive ion channels there are several other classes of ion channels present in cell membranes. The other classes of membrane ion channels are always closed unless they receive a signal, such as a change in transmembrane potential, telling them to open. It is the orchestrated opening and closing of the signal-requiring channels that allows electrical impulses to move along the length of a cell membrane. And, it is electrical membrane signaling that sets the stage for cell to cell communication.

Generally, signal-requiring ion channels only stay open for a short period of time after receiving the open signal. The signal to open can be a change in the local transmembrane potential, or the presence of a chemical that the channel recognizes as a signal, or a mechanical local stretching of the cell membrane. Each class of signal-requiring ion channel responds to only one type of signal. Such channels are said to be *gated channels*. Gated channels are either *voltage gated channels*, or *chemically gated channels*, or *mechanically gated channels*.

Regardless of when or where gated channels open, once they are open the same chemical and electrical forces that drive ions through the passive channels drive ions through these channels. The number of ions that flow through a gated ion channel is dictated by the channel's ion specificity and the sum of the local chemical and electrical forces on that ion.

Probably the best cell for illustrating how manipulation of these channels can move a signal along a membrane is the nerve cell. Nerve cells are most interesting cells. They are part of a group known as *excitable cells*. Other members of the group are the muscle cells – smooth muscle, skeletal muscle, and cardiac muscle.

The membrane of nerve cells has a shape that is very different than that of most cells. The membrane surrounds a fat body of intracellular fluid with many short spiny extensions plus one long extension. The short spiny extensions are called dendrites, and the one long extension is called an axon. The dendrites are where other neurons can attach with their own axons. Terminal ends of nerve cell axons can also attach to all types of muscle cells. In some areas of the human body nerve axons are short, but the axon of a single motor nerve to skeletal muscle can be as long as 3 feet.

The chemically gated ion channels and voltage gated ion channels described above are strategically located on nerve cell membranes. The distribution and ion specificity of such channels varies by type of nerve. A great deal of research has been devoted to determining which ion channels are located where on nerve membranes and under what circumstances their gates will activate.

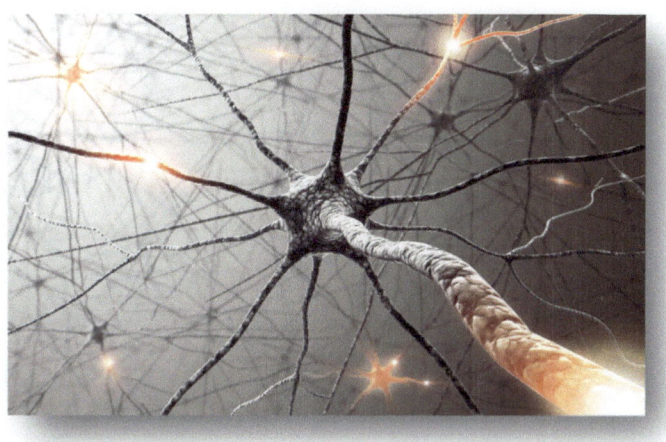

Figure 19: An artist's depiction of a group of nerve cells in action. Photo © 2012 ktsdesign

It is a common property that the terminal end of a nerve axon releases packets of a chemical called a neurotransmitter. Neurons are grouped by the type of neurotransmitter that their axon terminal releases. Neurotransmitter release happens only when the transmembrane voltage of an axon's terminal reaches a critical level.

A triggering transmembrane voltage for release of neurotransmitter is always of a different magnitude than the cell's resting potential described above. To change the magnitude of the transmembrane potential from the resting level at the axon terminal, it is necessary that gated ion channels be present. In addition to the channels being present, a signal must arrive that can open them. Theoretically the signal may be a divergence of the transmembrane potential at the axon terminal membrane, or a chemical that is recognized by an ion channel on the terminal membrane, or mechanical stretch of the terminal membrane itself in the vicinity of a channel.

In reality the array of ion channels in nerve axon terminals, depending upon where the nerve is located and its network pattern with other nerves, may include voltage gated channels, chemically gated channels, mechanically gated channels or combinations of these three. So, for simplicity we will continue this discussion with an illustration of a single nerve type, the skeletal muscle motor neuron.

How do muscle motor nerves use gated ion channels?

The cell bodies of the motor nerves that cause skeletal muscle to shorten reside in the spinal cord. Axon terminals of spinal nerves attach, or synapse, on dendrites of the motor nerves. A schematic drawing of a skeletal muscle motor neuron is pictured below. The membrane of dendrites of a motor nerve has chemically gated ion channels that respond to neurotransmitters released by the axon terminals of synapsing spinal nerves.

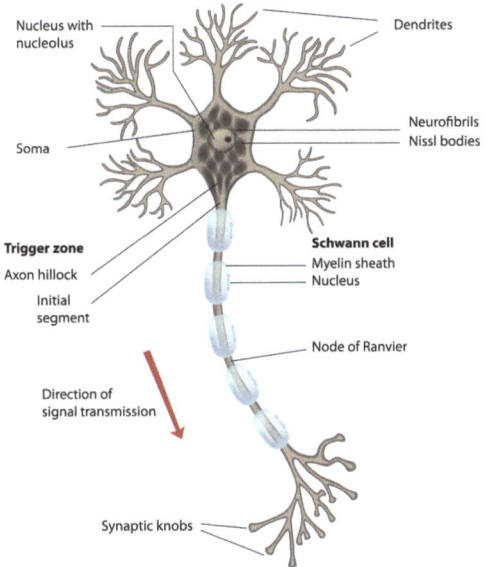

Figure 20: Schematic drawing of the components of a skeletal muscle motor neuron. Photo © 2012 Alila Medical Images

The axon terminals in the spinal cord synapsing on motor nerve dendrites are of various nerve types. Some of the neurotransmitters released may open ion channels that make the transmembrane potential of the motor nerve cell body *more negative – more polarized*. Other neurotransmitters released at the dendrites may open ion channels that make the motor nerve cell body's transmembrane potential *less negative/more positive – depolarized*. The net effect of the flow of ions through chemically gated channels on the dendrite membrane determines the transmembrane potential of the nerve cell body.

When the transmembrane potential of a motor nerve's cell body is depolarized to a critical level called the depolarization threshold, it causes voltage gated Na^+ channels in the nerve axon nearest the cell body to open. The beginning of the axon near the cell body, like the entire length of the axon, is populated with voltage gated Na^+ and voltage gated K^+ channels.

When voltage gated ion channels in the axon near the cell body open briefly they permit a flow of ions between intracellular and extracellular compartments down their concentration gradient. That flow of ions changes the transmembrane potential locally. The local change in transmembrane potential briefly opens more axon voltage gated channels a little further away from the nerve cell body.

The first axon channels to open when the nerve cell body reaches depolarization threshold are voltage gated Na^+ channels. Na^+ rapidly diffuses into the axon down its concentration gradient. Na^+ is also drawn by the negative charge on the inside of the membrane. With the entrance of Na^+ the transmembrane potential shifts from its resting negative value to a positive value. The positive transmembrane potential triggers closing of the Na^+ channels and opening of voltage gated K^+ channels. K^+ then diffuses out of the cell because of its concentration gradient and because it is repulsed by the positive nature of the inside of the cell membrane. The sudden loss of K^+ positive charges shifts the transmembrane potential back toward the resting membrane potential. The K^+ voltage gated channels close as the resting membrane potential is re-established. The time required for this process from opening of the Na+ voltage gated channels to complete repolarization of the cell membrane to its resting value is about 2 milliseconds.

During the period of time that Na^+ is entering the axon, some of this positive charge diffuses along the inside of the membrane away from the open channel. Nearby regions of the axon membrane then reach threshold depolarization opening more Na^+ channels, and the process repeats itself along the length of the axon.

The process of membrane depolarization only moves from cell body toward the axon terminal. This is because of the slow closing process for both the Na^+ and K^+ voltage gated channels. The closing process appears to be more complicated than the opening process and takes much longer. Until the closing process is complete, these channels will not reopen even if the transmembrane potential surrounding them remains positive.

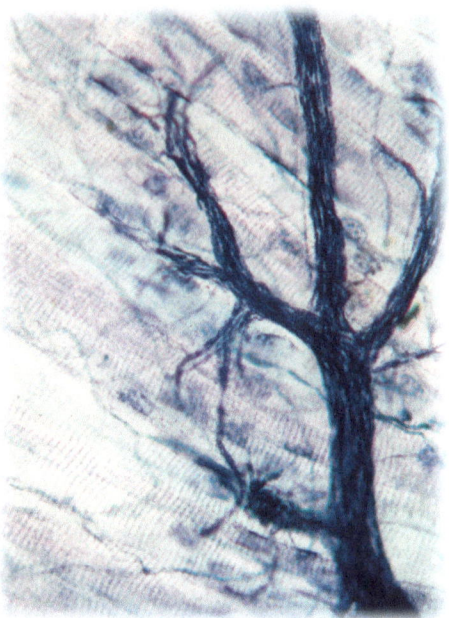

Figure 21: Photomicrograph of a motor nerve and its synapses on skeletal muscle cells (200x) Photo © 2012 Jubal Harshaw

This process repeats itself along the membrane of the axon altering the transmembrane potential sequentially all the way to the axon terminal. When the alteration in membrane potential reaches the axon terminal, it causes release of a neurotransmitter from the end of the motor nerve. Neurotransmitter released by skeletal muscle motor nerves is a chemical called acetylcholine.

Acetylcholine is recognized by chemically gated ion channels on the muscle cell membrane at the place where the axon of the motor nerve terminates.

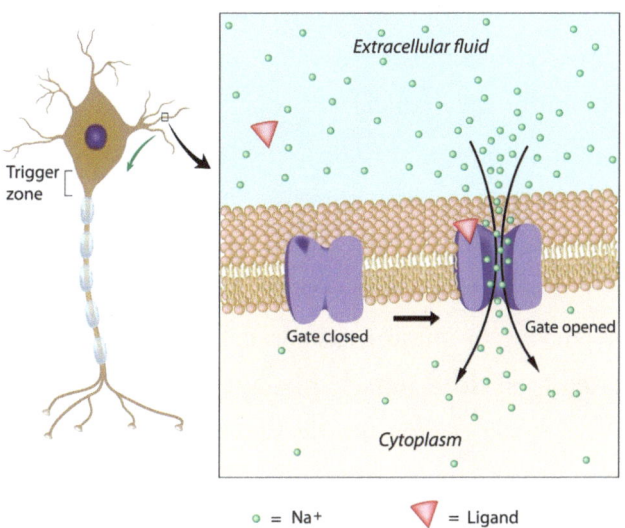

Figure 22: Model of a chemically gated membrane ion channel and a motor nerve, Photo © 2012 Alila Medical Images

Ion flow through the chemical gated channels of the muscle membrane alters muscle transmembrane potential. When there is a threshold depolarization of the muscle cell plasma membrane, nearby voltage gated channels along the muscle membrane briefly open.

The sequential activation of the voltage gated channels of muscle cell membrane is like the sequence along the nerve axon. Muscle cell membranes are unusual because they penetrate deep into the intracellular fluid compartment in a tubular fashion. Eventually the change in muscle cell transmembrane potential will reach the calcium ion storage areas deep inside the cell causing release of Ca^{++}. Release of stored calcium sets in motion the mechanical process of muscle cell contraction.

The motor neuron therefore integrates signals from spinal neurons that are in turn receiving signals from other neurons throughout the body. The information collected by the motor neuron cell body is then communicated to skeletal muscle through its long axon. All this long distance communication is accomplished simply by opening and closing of gated ion channels letting electrochemical properties of the ions and diffusion gradients between intracellular and extracellular fluid compartments accomplish the work.

Summary

To summarize there are several key concepts to remember from this chapter.

- ❖ Each of the fluid compartments of the body has a distinct character. Much chemical energy and anatomic creativity is invested in stabilization of fluid compartment components.

- An *electrical gradient* across a membrane separating two fluid compartments exists when one surface of the membrane has a positive charge and the other surface has a negative charge.
- The voltage difference between the inside and outside of a cell membrane is called a *transmembrane potential*.
- A *chemical gradient* exists when the concentration of a particular molecule differs in the fluid compartments on either side of a membrane.
- The term *electrochemical gradient* was devised to describe the combination of chemical and electrical forces that drive electrically charged potassium (K^+), sodium (Na^+), chloride (Cl^-) and calcium (Ca^{++}) through channels (pores) in cell membranes.
- Transmembrane potentials caused by leaky or 'passive' channels are said to be *equilibrium cell membrane potentials*.
- The combined effect of passive ion channel equilibrium potentials and ion pump activity creates a cell's *resting potential*.
- Beside passive ion channels there are several signal-requiring ion channels in cell mem-

branes. Such channels only open to an appropriate signal.
- ❖ Signal-requiring channels are said to be gated channels. Gated channels are either *voltage gated channels*, or *chemically gated channels*, or *mechanically gated channels*.
- ❖ Once gated channels are open the same chemical and electrical forces that move ions through passive channels move ions through these channels.
- ❖ Nerve cells serve as models for how ion gated channels in membranes separating intracellular fluid from intracellular fluid are useful for communication between distant parts of the body.

KEY WORD INDEX

CHAPTER 1
Energy, chemical, atoms, electrons, probability equations, orbital, hybrid orbital layer, molecule, molecular kinetic energy, chemical bond, covalent bond, low energy bond, high energy bond, ionic bond, molecular energy transfer, activation energy of a chemical reaction, enzyme

CHAPTER 2
Water molecule, molecular polarity, positive partial electrical charge, negative partial electrical charge, pH, hydronium ion, cation, anion, hydrogen bonds, surface tension of water, wetting a surface, hydrophilic, hydrophobic, moles, molar, buffer, acid, base, bicarbonate buffer

CHAPTER 3
Diffusion, osmosis, osmotic pressure, hydrostatic pressure, concentration gradient, semi-

permeable membrane, cell membrane, membrane water pore, membrane ion pore, pressure exerted by static fluid

CHAPTER 4
Gas laws, photosynthesis, respiration, ideal gas, gas pressure, Boyle's Law, Charles's Law, Gay-Lussac's Law, Combined Gas Law, gas partial pressure, Dalton's Law of Partial Pressures, gas flow, atmospheric partial pressures, Henry's Law

CHAPTER 5
Fluid compartments of human body, intracellular fluid (cytosol), extracellular fluid (interstitial fluid), blood, lymph, passive ion channel, electrical gradient, transmembrane potential, chemical gradient, electrochemical gradient, potassium ion channel, sodium ion channel, equilibrium cell membrane potential, resting potential, voltage gated channel, chemically gated channel, mechanically gated channel, excitable cell, nerve, skeletal muscle motor neuron, dendrite, nerve cell body, axon, nerve terminals, depolarization

ABOUT THE AUTHOR

Dr. Margaret Thompson Reece has a Ph.D. in Physiology awarded by the University of California. She began her research career in 1981 at the San Diego Zoo.

Her teaching experiences include college students from freshman level to Ph.D. She has been an instructor for courses in physiology, anatomy, human genetics, biochemistry, and cell biology.

In addition to her experience in teaching, she has been a scientist in academic medicine and a CSO for Serometrix LLC, a biotechnology company.

To find more strategies for studying physiology visit her website at http://www.medicalsciencenavigator.com

www.ingramcontent.com/pod-product-compliance
Lightning Source LLC
Chambersburg PA
CBHW041059180526
45172CB00001B/31